Biomolecular Electronics

Biomolecular Electronics

Bioelectronics and the Electrical Control of Biological Systems and Reactions

Paolo Facci

ELSEVIER

AMSTERDAM • BOSTON • HEIDELBERG • LONDON
NEW YORK • OXFORD • PARIS • SAN DIEGO
SAN FRANCISCO • SINGAPORE • SYDNEY • TOKYO

William Andrew is an Imprint of Elsevier

William Andrew is an Imprint of Elsevier
The Boulevard, Langford Lane, Kidlington, Oxford, OX5 1GB, UK
225 Wyman Street, Waltham, MA 02451, USA

British Library Cataloguing in Publication Data
A catalogue record for this book is available from the British Library

Library of Congress Cataloging-in-Publication Data
A catalog record for this book is available from the Library of Congress

ISBN: 978-1-4557-3142-8

For information on all William Andrew publications
visit our web site at store.elsevier.com

Typeset by TNQ Books and Journals
www.tnq.co.in

Printed and bound by CPI Group (UK) Ltd, Croydon, CR0 4YY

14 15 16 17 18   10 9 8 7 6 5 4 3 2 1

To Bianca, Maria, and Benedetta

Una lux illuminat omnia
(G. Bruno, Sigillus Sigillorum)

Contents

Preface

At the beginning of the 1990s biomolecular electronics asserted itself as a powerful novel approach to combine the most advanced technology humankind has ever developed with the most sophisticated type of matter known in the Universe, biological matter. Expressions such as self-assembly, specific recognition, natural evolution, and so on, suddenly became very popular and widely used (and sometimes, abused), including by people lacking a rigorous biological education.

I think that words (and the concepts that these words signify) do not belong to anyone in particular and everybody has the right to use them at will. Sometimes, this can turn out to be a little annoying for educated people but more often, especially in science, it helps to provide an undeniably advantageous change of paradigm that can foster cultural advancement. Apparently something similar has happened also in the development of biomolecular electronics. Cross-contamination among basic and applied disciplines helped a lot to shape the brand new research field that aimed at selecting and exploiting molecules of biological origin for implementing novel hybrid devices. These last were characterized by a novel concept according to which the molecule itself constituted the device, at variance with the classical top-down approach borrowed from fabrication processes used in solid-state electronics.

In this regard, the concomitant advent of nanotechnology resulted in a powerful boost for biomolecular electronics, which naturally faced the issue of connecting electronic circuits and leads to single molecules.

After more than 20 years from its birth one must admit, however, that real practical biomolecular electronics applications, in terms of operating devices on the market, are still dramatically missing and the legitimate doubt arises that they will never appear!

This outcome is indeed quite common in applied science and often accompanies those cases that were characterized by a generalized over-excitement about their expectations. To date, not many experts trust in the future advent of transistors made from proteins or DNA.

At any rate the positive side-products, brought about by more than two decades of research in this field, are numerous and range from a deep understanding of electron transport through single biomolecules, to the elucidation of the key role of contacts with metal leads and that of redox electronic levels in assisting electron transport, to the correct appreciation of conformational variations in modulating molecular transport properties. The essential role of water in enabling and preserving functional molecular conformations has once more been clarified.

I would list in the positive outcomes of the research activity in biomolecular electronics also the realization that its true added value is not likely to be in competing with much better-established technological paradigms to replace already existing and well performing devices (e.g., nanotransistors); rather, it is in having developed a series of approaches suitable for the effective interfacing of biological molecules and reactions with conventional electronic systems.

Indeed, this fact enables electronic control over the functional activity of biological molecules (e.g., metalloenzymes) and related reactions and paves the way to a future scenario where biological phenomena and systems will be controllable and driven technologically by external electric signals. The first steps towards this novel, charming scientific and technological adventure are what the present book is about.

During my activity in the field of biomolecular electronics I have had the opportunity of sharing with numerous young collaborators, students, and colleagues many ideas, efforts, lack of success, and, sometimes, also the happiness of exalting results. I am grateful to all of them. In particular, I thank Andrea Alessandrini, Victor Erokhin, Gerard Canters, Lorenzo Berti, Marco Salerno, Dario Alliata, Laura Andolfi, Carlo Augusto Bortolotti, Mimmo Gerunda, Paolo Petrangolini, Marialuisa Caiazzo, Elena Angeli, and Loredana Longo, who more than others have collaborated with me in this field and without whom I could not have reached the present view of this subject.

Paolo Facci, Genova

CHAPTER

Biomolecular Electronics

1

CHAPTER OUTLINE

1.0 What is biomolecular electronics?

With the expression "biomolecular electronics" we refer nowadays to that branch of technology that exploits molecules or molecular systems of biological origin for interaction with modern electronics.

The idea of using biomolecules to assemble hybrid electronic devices stems from molecular electronics (Aviram & Ratner, 1974) and, as such, is intimately connected with the advent of nanosciences and nanotechnologies, dating back to the beginning of the last decade of the past century. The interest in biomolecules stems not only from their size, which is typically in the nanometer range (at least in two spatial dimensions). Rather, it is often connected to a particular standpoint from which one can regard them.

Indeed, an interesting point of view is that of regarding biomolecules in general and proteins in particular as self-contained, nanometer-sized functional units that are highly specialized and efficient in performing a certain functional task. Their efficiency in performing a particular activity is traced back to the fact that biomolecules, being parts of living beings, have naturally evolved over billions of years, reaching the current "optimal" degree of specialization for accomplishing a given task.

One can often come across this opinion reading the specialized literature or attending topical conferences; however, it appears quite questionable in light of a slightly less naïve understanding of the Theory of Evolution (Fodor & Piattelli-Palmarini, 2010); nevertheless, it is perhaps a good enough starting point to understand the historical motivations which led to the

Biomolecular Electronics. http://dx.doi.org/10.1016/B978-1-4557-3142-8.00001-7

remarkable interest shown by a sizable part of the interdisciplinary scientific community in the use of biomolecules for assembling electronic devices.

Another element motivating the interest in proteins as components of electronic circuits is their sizes. Indeed, these molecules often configure self-consistent, functional units with typical sizes in the nanometer range. Therefore, it is conceivable to imagine functional devices as small as a single molecule, especially in view of the fantastic progresses in miniaturization and parallelization made with nano-lithographic techniques.

Furthermore, the rich chemistry that particularly characterizes protein-forming monomers, i.e., the 20 natural amino acids, makes these molecules amenable to various kinds of chemical functionalization, thus improving their reactivity towards desired functional groups and enabling, as a consequence, their ready chemisorption on pre-functionalized structures and surfaces. Moreover, the exceptionally fine-tuned recognition properties of biological structures (e.g., complementary nucleic acid strands, DNA-binding proteins, antibody-antigen or receptor-ligand pairs, etc.) could allow for self-assembling of more complex structures in the nanometer-sized gap between meso/nanoscale electrodes.

Among the various classes of molecules of biological origin, two of them have been especially focused upon by researchers in the field: DNA and proteins. The reasons for this choice are different; in the case of DNA, its robustness and ease of handling are relevant factors. Nevertheless, a further important aspect fostering DNA as an interesting molecule in biomolecular electronics stemmed from the idea that, if we could understand the mechanism by which electrons are driven along the double helix during DNA repair of damage caused by oxidative stress, we would obtain, as a byproduct, a way to produce molecular wires, that is, one-molecule-thick conducting wires, to be used in assembling (bio)molecular circuitry. If DNA can be regarded in terms of molecular wire, what else does one need to assemble molecular circuits? Some functionally relevant electronic elements, of course. Those were identified as the proteins, which are indeed deputed to perform the main functional tasks in the organisms that express them.

At this point a relevant question arose that has not been completely answered yet, about the electrical conductivity of biomolecules. It is indeed obvious that at the basis of any electronic circuit there is a need to deal with wires and elements that can conduct or, more generally, interact functionally with electron flows, i.e., electron currents.

Whereas this issue is still open in the case of DNA, which displays good conductivity over short ranges (of the order of nanometers) and appears to be an insulator over larger distances (Guo et al., 2008), the situation regarding proteins is more complex and requires detailed analysis because of the large variability of their properties (mirroring the differences in protein structures).

In the case of DNA, its poor conductivity, at least in the mesoscopic range, has prompted various approaches for enhancing it, based on doping (with metal ions or intercalating agents (Ban et al., 2009)) or on the use of DNA as a molecular template for metal nanowire growth (Berti et al., 2005). In the case of proteins, which possess an intrinsically higher chemical and structural variability compared with DNA, it is not surprising that specific transport proteins, carrying either electrons or ions, soon attracted the interest of researchers.

1.1 Proteins and biomolecular electronics

Typically, electron transport proteins are water-soluble molecules (rare exceptions apart; see, e.g., cytochrome b_{561}, a transmembrane three-heme cytochrome), whereas ion transport proteins are membrane molecules requiring specific environmental conditions (incorporation in a lipid bilayer) to be functionally active (see section 3.6.1). According to this difference, whereas the former are more easily implemented in electronic circuits, the latter are more demanding to exploit in a nano-circuit configuration. Ion channels have been used to gate transistors in hybrid devices thanks to their selectivity for specific ions (Bernards et al., 2006) and the inherent similarity of the gating mechanism of voltage-gated ion channels to the basic features of a field-effect transistor (Bezanilla, 2005). A specific application of ion channels in electronic circuits for sensing applications is the coupling of the channels reconstituted in lipid bilayers with other electronic nanostructures such as semiconductor nanowires (Misra et al., 2009). Examples of these applications are the possibility, by changing the pH of the solution, of controlling the conductance of a silicon nanowire covered with a lipid bilayer integrating peptide pores and the creation of nanobioelectronic transistors exploiting ion pumps such as Na^+/K^+-ATPase (Huang et al., 2010).

However, the most popular and charming class of proteins that seemed to be particularly suitable for implementing electronic devices is the redox metalloproteins. These very important molecules are indeed able to shuttle electrons between molecular partners by reversibly changing the oxidation state of the metal ions they contain (see sections 4.1, 4.1.1, and 4.1.2).

Following the route defined by conventional solid-state electronics, the research efforts in biomolecular electronics have generally focused on demonstrating transistor-like behavior that proteins might display when positioned between two electrodes. However, considering the classic definition of a transistor, we should stress here a different meaning for the same word. A transistor represents a device composed of at least three electrodes (source, drain, and gate) which is able to amplify or switch the current between a pair of electrodes (e.g., source-drain) by controlling the voltage or current between two of these three electrodes (e.g., source-gate or drain-gate). In conventional electronics the controlling signal is an electric signal. In the context of proteins, among the possible signals that can function for gating a hybrid device we include also other signals of a chemical or physical nature. Among them one has to consider all the parameters that are able to modify, at constant bias voltage, the transport properties of a protein or of a protein layer sandwiched between two electrodes. In a similar way, protein conformational variations can induce changes in its transport properties or in those of another nanostructure to which the protein is coupled. Among these signals one can list variation in environmental pH, temperature, substrate recognition by enzymes and so on. The sensitivity to any of these signals confers sensing properties to the related devices.

We will consider the main results obtained along the pathway of understanding the behavior of redox metalloproteins when used as active parts of hybrid electronic devices in Chapter 4. Here, we continue by noting that other features of proteins have also attracted researchers' attention and have been used in various attempts to exploit the properties of natural biomolecules to implement electronic devices. There is, for instance, the case of antibodies. These molecules are part of an organism's immune system and are known for their specific recognition and binding capabilities against the corresponding antigens. Their use has been recently demonstrated in assembling single-protein devices (Chen et al., 2012). Particularly, antibodies of the IgG type (immunogammaglobulin) have been raised against Au nanoparticles, 5 nm in diameter. An IgG molecule can bind up to two of these gold nanoparticles, one per F_{ab} fragment, where the specific recognition sites for antigens are located. This specific binding provides a stable bond. Gold nanoparticles are attached to e-beam lithography-fabricated 10-nm gap electrodes that act as source and drain electrodes. Once the electrodes have been decorated with nanoparticles, they are exposed to IgGs that specifically recognize the nanoparticles, binding them firmly. This results in a configuration that can be gated electrically, by means of a proximal electrode, and optically, by the action of a CdSe Qdot attached to the IgG molecule at its F_c fragment. While irradiated, the Qdot absorbs light, providing an induced dipole moment (confined exciton) that operates an effective gating of the single-protein transistor, probably mediated by conformational modifications occurring in the biomolecule. A richness of features is observed, including current values of the order of a few hundreds of picoamps, rectification, and negative differential resistance (NDR), which is ascribed to the presence of the F_c fragment. Such an exciting implementation is surprising because it shows it is possible for an electric current to flow through a large biomolecule that, physiologically, is not thought to have any electron transport capabilities.

This fact raises questions about protein conductivity and, hence, about the suitability of proteins as materials for electronics. Whereas theoretical considerations about protein conductivity and electron transfer in general will be dealt with later in this book (see Chapter 4), systematic experimental studies (Ron et al., 2010) suggest that electron transport through proteins, even measured in the solid state, is more effective than through a generic insulating material. There, the peculiar conductivity displayed by redox metalloproteins is confirmed, but even proteins without a physiological role in electron transfer (e.g., albumin, bacteriorhodopsin) display electron transport decay constants β in the range 0.12–0.27 $Å^{-1}$ that denote them as promising materials for bioelectronic applications. Indeed, with the same molecule, bovine serum albumin (BSA) assembled as monomolecular layer between source and drain electrodes of a vertical, 4-nm-gap hybrid transistor, a marked gate potential dependence as well as sizable currents have been obtained (Mentovich et al., 2009). BSA has also been used as an electronic material after proper doping with C_{60} molecules (Mentovich et al., 2012), highlighting once more the use of a protein matrix as a promising material for (self)assembling electronic devices.

As already stated, among the proteins that have been considered for bioelectronics purposes, a peculiar role is that played by redox metalloproteins, i.e., proteins containing one or more metal ions in their active centers that physiologically accomplish the task of transferring electrons between molecular partners. Metalloproteins represent about 30% of the entire proteome and the subset of the redox ones, including blue copper proteins, heme-based proteins, iron-sulfur complexes, molybdenum enzymes, and chlorophyll-bearing proteins, are specifically devoted to, or involve in their functioning, electron transfer. Indeed, all of these classes have been considered in bioelectronics applications. Among the most relevant attempts to involve redox metalloproteins in solid-state electronic devices, it is worth recalling implementations exploiting the blue copper metalloprotein azurin, which will be the subject of several studies also reported in Chapter 4.

Dealing with electronic transport properties studied at the level of a single or a few proteins, scanning probe microscopy-related techniques provide a useful experimental setting. In sections 4.5 and 4.5.1 we will concentrate on the particular technique of electrochemical scanning tunneling microscopy. Here, it is worth recalling that some studies involving single/few molecules by means of current probe atomic force microscopy (CPAFM) on the blue copper protein azurin (see section 4.1.1) have revealed an applied load dependence in the I-V curves, enabling a separation of the effects of distance and tunneling barrier variation in current changes upon increasing load (Zhao et al., 2004). Furthermore, load and bias voltage-dependent negative differential resistance (NDR) features suggested a role for the active site redox ion in mediating electron transport through the molecule (Davis et al., 2006, 2008).

1.2 Proteins and planar devices

The data retrieved from surface-immobilized redox metalloprotein samples by scanning probe techniques operated in the electrochemical environment (see Chapter 4) suggest clearly that these molecules behave like molecular switches, being able to allow or prevent electrons flowing through them according to the availability of molecular electronic levels in between the Fermi levels of tip and substrate, considering also the coupling of these levels with the environment. The possibility of switching the conduction state of an object such as a molecule by gating the flow of electrons through it represents indeed the basic feature of a molecular (nano)transistor (see Figure 4.21). Therefore, the idea of trying to implement a single (few) molecule(s) transistor exploiting the electron conduction properties of metalloproteins (e.g., azurin) arises naturally (Facci, 2002). Redox metalloproteins arranged in one-molecule-thick films in a gap defined by a pair of planar electrodes could hence act as a channel in a field-effect three-terminal device. Of course, the problems involved in implementing such a device immediately appear to be very serious and numerous. Among them, it is worth stressing two of the most relevant ones, which are: (i) the implementation and theoretical description of effective electrical contacts between metalloproteins and metal electrodes, and (ii) the understanding and optimization of the mechanisms and conditions of intermolecular electron transfer in 2D ensembles of metalloproteins, considering also the fact that many of these devices are implemented in water-less conditions (only tightly bound water molecules are present). Here, we will not discuss in detail the *necessary conditions* for the construction of these devices, i.e., the strategies for protein immobilization on a surface. We refer the reader interested in this topic to reviews in the literature (Willner & Katz, 2000; Lösche, 1997; Ulman, 2001). Albeit a final solution for these classes of problems is still lacking, the practical realization of working devices is possible and it has been accomplished in fact. Indeed, it has been shown that proteins in a dry environment have an electronic conductivity comparable to that of conjugated molecules, highlighting the fact that proteins offer a very efficient medium for electron transport (see section 4.4.1). It is typically assumed that electron transport between two electrodes separated by a monomolecular protein layer has an exponential dependence on the tunneling barrier width (the thickness of the protein layer can be used as a good approximation for the barrier width). The decay constant can be exploited to investigate the type of transport mechanism, specifically to distinguish between a superexchange or a hopping mechanism (Ron et al., 2010). In fact, it is in principle possible to distinguish between a direct non-resonant tunneling and a multistep tunneling process (Ron et al., 2010). It is also possible that the characterization of these devices could provide information on the basic transfer properties of proteins even if they are studied in an environment different from that in which they perform their natural task.

The implementation of a Field Effect Transistor (FET) – or "single particle transistor-like protein device" – should take advantage of state-of-the-art lithographic techniques for the definition of planar (nano)electrodes. Figure 4.21 depicts the operating principle of a generic single metalloprotein planar transistor. Here, a redox metalloprotein is located in the nanometer-sized gap between two planar electrodes and is electrostatically coupled to a gate electrode. The coupling is responsible for shifting the electronic levels of the molecule with respect to the Fermi levels of the metal leads, enabling or hindering the electronic flow via the aforementioned levels. In this case, the choice of using metalloproteins is dictated by the possibility of exploiting energy levels that are specific to redox molecules. The idea is to use the typical features observed in electrochemical experiments on these molecules (see section 4.4.1) to perform a task in the context of solid-state electronics.

To date, a limited number of different planar nanoelectronic devices have been implemented. They typically make use of those redox metalloproteins, which have so far shown the most robust performance in withstanding large electric fields without undergoing denaturation and in surviving rather non physiological environments such as that of the surface of a hybrid electronic device (operated in a water-less environment).

The implementations reported so far have been for two- and three-terminal devices; they have had the aim of demonstrating the feasibility of such hybrid devices and have been used especially to characterize those particular systems, trying to shed light on the role of molecular organization and on that of the particular metal ions in the protein.

The first implementation of a hybrid nanoelectronic device was of a solid-state molecular rectifier based on a (sub)monolayer of azurin molecules (Rinaldi et al., 2002). In this case, the main interest was in the transport properties of the protein layer. A monomolecular carpet of metalloproteins was chemisorbed on the surface of thermally grown SiO_2 in between two gold nanoelectrodes (Cr-Au) defined by e-beam lithography (EBL). The chemistry used for the chemisorption of the protein carpet was based on silanes. The construct guaranteed a covalent immobilization of the molecules on the surface as well as a defined uniform protein orientation. Control experiments on non-uniformly oriented azurin hybrid devices were performed using another immobilization strategy. The achievement of a uniformly oriented protein film in between nanoelectrodes is of extreme importance for the performance of a hybrid device for two main reasons: (i) it enables a higher surface coverage, which can facilitate intermolecular electron transfer between neighboring metalloproteins; (ii) for a given protein coverage, it endows the molecular carpet with a uniformity in the location and orientation of the Cu-containing redox active sites, increasing the probability of electron transfer. Another important aspect, which is strictly related to protein orientation at the surface, is protein electrostatics. Proteins, being complex zwitterionic molecules, possess on their surface a variety of charged and polar groups whose global effect is described by MEPs (molecular electrostatic potentials). MEPs are recognized to play a major role in protein interaction and bio-recognition in solution and drive the phenomena of self-assembling at surfaces and adsorption kinetics. MEPs for azurin have been computed in a range of different conditions involving different protein oxidation states, solution pH and ionic strength. Calculations have shown that azurin, in the experimental conditions used in the work being discussed, possesses a strong dipolar moment (150 D) that can sum up in an oriented, 2D molecular assembly, influencing the overall field sensed by electrons flowing in the monomolecular layer.

The electrical characterization of a two-electrode device with a 60 nm gap has been performed in the case of uniformly versus randomly oriented protein immobilization. Some important features and differences were found: (i) the uniformly oriented sample showed a marked rectifying behavior (rectification ratio of 175 at 1.5 V vs a ratio of just 10 for the randomly oriented counterpart); (ii) the uniformly oriented sample showed a 10-fold higher current with respect to the randomly oriented one; this effect, along with that of rectification, can be attributed to the molecular dipoles summing-up in the self-assembled edifice and providing an electrostatic field superimposing on the external bias; (iii) marked steps in the I-V curves, tentatively ascribable to the involvement in the conduction mechanism of protein redox levels, appeared in both samples, albeit different in intensity. This first demonstration of a metalloprotein-based molecular rectifier showed poor stability to aging and to environmental conditions, suggesting, however, that the nature of the observed phenomena was due to the presence of the molecules.

Another work addressed several open issues related to protein sample purity, and the role of the Cu ion in the azurin active site (Rinaldi et al., 2003). In nanogaps 50 to 100 nm wide, it turned out that recombinant native azurin performed better that the wild type since it showed a rectifying ratio of 500 at 10 V. Moreover, an extensive study with different molecules (recombinant native, Zn-azurin, and apo-azurin – without the metal ion), all immobilized in a uniform orientation, was undertaken in order to understand the role of the particular metal ion in the active site, in a similar fashion to what was previously studied by EC-STM at the level of the single molecule (Facci et al., 2001; Alessandrini et al., 2003). These results confirmed the key role of Cu ions and that of the electronic levels brought about by ion presence, in assisting current flow through the molecular carpet. Zn-azurin and apo-azurin, albeit almost identical to native Cu-azurin in molecular structure, were in fact unable to let any appreciable current flow in the device. Finally, the role of relative humidity in preserving the performance of the hybrid device over time was revealed. A relative humidity of around 50% was found to be optimal for device performance and durability.

As a further step, a three-electrode device has been implemented (Maruccio et al., 2005), i.e., a device where the current through the molecular carpet was controlled by a gate electrode in a FET-like implementation. In this particular case, attention was devoted to the possibility of finding in the electron transport phenomenon features that could resemble a transistor-like effect identified in EC-STM experiments (see section 4.6) and related to the redox properties of the proteins. The experimental configuration for such a device was pretty much the same as the two-electrode one (EBL defined Cr/Au arrow-shaped nanoelectrodes facing a 100 nm gap on a Si/SiO_2 substrate). However, the main difference was the presence of an Ag electrode on the back of the structure, making an ohmic contact with the p-doped Si substrate. This electrode was used as gate in a FET configuration. The main findings of this work were the presence of a marked resonance in the measured transcharacteristics, which displayed a peak-to-valley ratio of 2 at V_g= 1.25 V. Only devices assembled from Cu-azurin displayed such a feature, at variance with Zn- and apo-azurin.

Interestingly, the switching behavior of this protein-based FET was also modeled in a simplified frame of protein chains of alternating reduced and oxidized molecules, invoking a hopping mechanism for intermolecular electron transfer. Within this framework, it was possible to account for the main experimental features of the implemented device, namely: (i) the appearance of the described resonance in transcharacteristics, and (ii) the marked onset in the I-V curves, as already described in the implementation of the two-terminal devices.

Recently, Richter et al. developed a vertical transistor configuration based on a small channel in which proteins were hosted (Mentovich et al., 2009). In this configuration, a central gate electrode was exploited to activate a self-assembled monolayer of proteins sandwiched between two metal electrodes representing the drain and source electrodes. It is to be stressed that, in contrast to the previously described planar devices, in this configuration the current flow was in the same direction as the main axis of the self-assembled proteins, whereas in the previous case the current was perpendicular to the main axes and, moreover, in the set-up by Richter et al., the current was measured over a very small monolayer area. As a consequence, in the former set-up the intermolecular electron transport mechanism had a relevance, which was not present in the latter one, where the main transport occurs across just one protein. It has been shown that the gate potential is able to align the energy levels of the proteins inside the window defined by the Fermi levels of the source and the drain. By this technique it has been possible to ascertain the connection between the redox properties of the proteins and the presence of a negative differential resistance feature in the I-V curves (Mentovich et al., 2011). Once again, the azurin protein was used to perform this experiment and the results have been compared to the case of the apo-azurin protein. The absence of the NDR peak in the case of apo-azurin suggests that the observed feature is related to the redox center of azurin.

1.3 The future of biomolecular electronics

After more than two decades of intense activity in the field, it is time to ask ourselves whether this charming discipline is likely to have a future as an effective competitor for conventional electronics or whether its achievements, albeit interesting, are to be considered just proofs of principle and will never get to the level required for representing a real alternative to the current technology.

This is of course a matter of debate, and it is a waste of time to try to elaborate an answer that is largely shared among the scientist active in the field. Rather, it is perhaps more useful to present some considerations that can contribute to the development of one's own judgment on the subject.

The possibility that a novel technology, relying on approaches so different from those of the current solid-state-electronics-based one, has any chance of replacing or even only supplementing the existing one appears indeed quite unlikely. Several reasons seem to suggest this conclusion; probably the main one is that in such a technologically-intensive field as the modern electronic industry, even the tiniest modification of the production processes requires such a huge economic effort that it is usually unbearable by single companies. In light of this fact, a radical change of perspective like that required by biomolecular electronics (which, for instance, would need to change the fabrication approach from top-down to bottom-up) would represent a genuine revolution, incompatible with the natural development of the market. Even the often-mentioned argument that optical lithography is rapidly approaching its physical limit dictated by the laws of diffraction applied to a radiation of finite wavelength, and that the size of the tiniest features could not decrease further, is regularly falsified by technological advances (to date, the tiniest transistor in production, belonging to the family "Ivy Bridge", has a size of 22 nm). The recent achievement of carbon nanotube-based transistors, whose production is compatible with the technological infrastructure used for producing CMOS commercial devices, provides a much more reliable insight into the likely near future development of electronics technology (Park et al., 2012).

Indeed, the bottom-up approach required for self-assembling biological macromolecules, despite being fascinating, does not provide the reliability required for massive production of identical devices. Using it for device production would require a different capability for rejecting defective pathways in integrated circuits, possibly operating in a dynamic way, thus recalling the behavior of real neural networks, which is, unfortunately, still a long way off for human technology.

A further major problem is then represented by the requirement for a water-based environment for a proper functioning of biomolecules (see section 3.5). This difficulty sounds to date almost unsurmountable, in light of the drastic incompatibility that both fabrication processes and device operation have with a wet environment!

Another obvious concern is connected to the robustness and stability (time, temperature, environmental changes, etc.) of a chip made of biomolecular devices. Even if there exists evidence for the increased thermal stability of biomolecules operated as part of an electronic circuit and shelf life can be made longer than that of the corresponding biomolecules in vivo or freely diffusing in solution, the comparison with solid-state inorganic devices appears frankly unbearable.

The current levels sustained by biomolecular devices are another important cause of concern. It is impossible to foresee the implementation of any biomolecular device operating at high power.

Last but not least, the best performance dealing with switching ratios obtained with biomolecular electronic devices is by far inferior to that of any, even modest, solid-state device.

Perhaps, the most promising and realistic possible application niche where biomolecular devices can strive to compete with solid-state counterparts is that of (bio)sensing. Indeed, the richness in surface chemistry and the intrinsic potential of many biomolecules to undergo conformational changes upon environmental parameter variation and/or binding of molecular effectors make them ideally suited for sensing even the slightest changes, when arranged in single/few-molecule devices as is the case in a bioFET (see for example, section 4.6). It is conceivable that such devices would be integrated with conventional solid-state electronics, providing just a compatible signal for further standard processing and elaboration.

At any rate, the general outcome of all these considerations suggests a remarkable reshaping of the extent and potentials of biomolecular electronic devices. Does it mean that there is no future for interfacing biomolecular systems with man-made technology? We believe that the answer to this question is no. Let us see why.

1.4 A novel idea: electrical control of biomolecular systems

In most cases, extensive efforts in any field of science and technology bring about relevant byproducts that are largely independent of the success of the original enterprise. It is not uncommon that relevant discoveries and applications come about from research aiming at totally different results (see, for instance, the biography of Irvin Langmuir reported in Tanford, 2004). The awareness developed during more than two decades of research in the field of biomolecular electronics puts researchers active in the field in the position of being able to master many aspects related to interfacing electronic systems and biological ones. Examples include the interplay between surface functionalization and molecular biology, which allows researchers to immobilize a given biomolecule in a desired site with a chosen orientation (Alessandrini et al., 2008); the availability of powerful, single molecule investigation and manipulation techniques such as scanning probe microscopes (Alessandrini & Facci, 2005); the deep understanding of the role of contacts with electrodes in measuring the electrical properties of single molecules; the role of solid-state surfaces in affecting biomolecular conformation and stability, just to quote some of the most relevant aspects.

On top of these, there is a further fundamental aspect that can be easily recognized as being at the basis of biomolecular electronics: that is the role that electrical phenomena play in biological systems. This is of course a major issue, chemical-physical in nature, that involves nearly all biological phenomena. It is enough to think of the role that electrical interactions play in maintaining the various molecular components of a biological system in solution, preventing their flocculation, the otherwise unavoidable fate of any colloidal system subjected only to Van der Waals-type forces (Israelachvili, 2005).

This pervasive role of electrostatic interactions in biological systems is connected to the properties of the chemical moieties forming the various biopolymers, which are often zwitterionic or charged (Branden & Tooze, 1999), as well as to the presence of a polar solvent that dissociates salts dissolved in it.

The role of electrical interaction, however, affects a very large number of biological phenomena in a much more detailed and refined way. Just to quote some examples, we can think of the role of ions such as Na^+ and K^+ in generating and maintaining transmembrane potential across the plasma membrane; the effect of charged residues on the conformation and conformational modifications of enzymes, antibodies, and many other proteins; the peculiar, albeit fundamental, function that positively charged residues (e.g., arginines) have in enabling the response of voltage-gated ion channels to changes in transmembrane potential (see section 3.6.1); the major role that biological redox reactions play in nearly all the key metabolic, energetic, and control phenomena in cells; the effect that modulation of surface charges has in controlling cell aggregation (e.g., human erythrocytes possess a net negative charge, $\cong 2.5 \cdot 10^{-12}$ C, on their surface that prevents them from clumping together; treating cells with a proteolytic enzyme that cleaves the charged residues on their surface proteins causes red blood cell clotting).

At the systemic level too, biological matter is governed by electrical interactions and phenomena and relevant examples are really countless. Just considering the activity of neurons and the nervous system in general, or the behavior of cardiomyocytes or muscles, one has a glimpse into the pervasive role that these phenomena play.

Of course, such a major role of electrostatic interaction in nearly all the structural and functional aspects of biology is also routinely exploited by technology in many contexts. The charges brought about by biological macromolecules, for instance, are used to separate biomolecules according to their mobility in an electric field when one performs gel electrophoresis. Biologists take advantage of electric charges for sorting cells in flow cytometers, and doctors apply electric shocks across patients' chests in the case of heart attacks or arrhythmias to reestablish the normal rhythm.

Whereas there is no point in trying to list all the relevant cases in which electrical interactions play a role when considering biological matter and phenomena, it is instead very important to reflect on the potential relationship that exists between these phenomena and our ability to control them.

This is really a point of key relevance for this book, since all the rest of the present work will stem from the considerations we will develop here.

Indeed, on the one hand we are now quite convinced of the relevance and ubiquity of electrical phenomena in biological matter (the most sophisticated one in the known Universe) and systems. On the other hand, we are constantly exposed to the most advanced technology that humankind has ever developed, which is electronics. From the research activity in biomolecular electronics we have learnt a lot about how to handle biological entities and how to interface them with the electronic environment.

It is therefore possible to make a further step towards the synthesis of these two realms, biology and electronics, to come to a new level of interaction between them. In other words we are about to enter the charming field of extending technological (electrical) control over biological phenomena and systems.

This control, as will be thoroughly exposed in the following chapters, will exploit the peculiar features characterizing the handling of electric fields in a water-based salty environment, referring therefore to the concepts of electrochemistry, and will be extended to treating phenomena that, so far, have not been considered amenable to direct electrical control, such as the modulation of gene expression level in bacteria (see Chapter 7) or the extraction of energy from the photosynthetic cycle in plants (see Chapter 8).

This novel concept aims at identifying those biological systems that, more than others, appear to be amenable to a seamless integration with electronics (e.g., power supplies, amplifiers, potentiostats, etc.) and have the potential to give rise to hybrid systems where technological control of a particular aspect such as the oxidation state of certain molecules can drive and rule the behavior of a biological system in a desired direction.

We will have the opportunity to realize how large is the extent of phenomena that can in principle be subjected to external electrical control, ranging far beyond the simplest paradigmatic examples of photosynthetic bacteria and organelles and involving, for example, stem cell differentiation, modulation of tumor suppressor (e.g., p53) activity, and electrical blood pressure control (see chapter 8).

In the next chapters the reader will be able to appreciate how the heritage left by biomolecular electronics in many relevant scientific and technological contexts can really boost the novel field presented here towards a future that we foresee as bright and rich in potentially charming consequences.

1.5 References

Alessandrini, A., & Facci, P. (2005). AFM: a versatile tool in biophysics. *Meas. Sci. Technol., 16*, R65–R92.

Alessandrini, A., Gerunda, M., Canters, G. W., Verbeet, M. P., & Facci, P. (2003). Electron tunnelling through azurin is mediated by the active site Cu ion. *Chem. Phys. Lett., 376*, 625–630.

Alessandrini, A., Berti, L., Gazzadi, G. C., & Facci, P. (2008). Imparting chemical specificity to nanometer-spaced electrodes. *Nanotechnology, 19*(355303), 1–5.

Aviram, A., & Ratner, M. A. (1974). Molecular rectifiers. *Chem. Phys. Lett., 29*, 277–283.

Ban, G., Dong, R., Li, K., Han, H., & Yan, H. (2009). Study on the electric conductivity of Ag-doped DNA in transverse direction. *Nanoscale Res. Lett., 4*, 321–326.

Bernards, D. A., Malliaras, G. G., Toombes, G. E. S., & Gruner, S. M. (2006). Gating of an organic transistor through a bilayer lipid membrane with ion channels. *Appl. Phys. Lett., 89*(053505), 1–3.

Berti, L., Alessandrini, A., & Facci, P. (2005). DNA-templated photoinduced silver deposition. *J. Am. Chem. Soc., 127*, 11216–11217.

Bezanilla, F. (2005). Voltage-gated ion channels. *IEEE Trans. Nanobiosci., 4*, 34–48.

Branden, C., & Tooze, J. (1999). *Introduction to protein structure* (2nd ed.). New York: Garland.

Chen, Y. S., Hong, M. Y., & Huang, G. S. (2012). A protein transistor made of an antibody molecule and two gold nanoparticles. *Nat. Nanotechnol., 7*, 197–203.

Davis, J. J., Wang, N., Morgan, A., Zhang, T., & Zhao, J. (2006). Metalloprotein tunnel junctions: compressional modulation of barrier height and transport mechanism. *Faraday Discuss., 131*, 167–179.

Davis, J. J., Peters, B., & Xi, W. (2008). Force modulation and electrochemical gating of conductance in cytochrome. *J. Phys.: Condens. Matter, 20*(374123), 1–9.

Facci, P. (2002). Single metalloproteins at work: towards a single-protein transistor. In T. Chakraborty, F. Peeters, & U. Sivan (Eds.), *Nano-physics & bio-electronics: A new Odyssey*. Amsterdam: Elsevier.

Facci, P., Alliata, D., & Cannistraro, S. (2001). Potential-induced resonant tunneling through a redox metalloprotein investigated by electrochemical scanning probe microscopy. *Ultamicroscopy, 89*, 291–298.

Fodor, J., & Piattelli-Palmarini, M. (2010). *What Darwin got wrong* (1st ed.). New York: Farrar, Strauss and Giroux.

Guo, X., Gorodetsky, A. A., Hone, J., Barton, J. K., & Nuckolls, C. (2008). Conductivity of a single DNA duplex bridging a carbon nanotube gap. *Nat. Nanotechnol., 3*, 163–167.

Huang, S. C., Artyukhin, A. B., Misra, N., Martinez, J. A., Stroeve, P. A., Grigoropoulos, C. P., Ju, J. W., & Noy, A. (2010). Carbon nanotube transistor controlled by a biological ion pump gate. *Nano Lett., 10*, 1812–1816.

Israelachvili, J. N. (2005). *Intermolecular and surface forces* (3rd ed.). New York: Academic Press.

Lösche, M. (1997). Protein monolayers at interfaces. *Curr. Opin. Solid State Mater. Sci., 2*, 546–556.

Maruccio, G., Biasco, A., Visconti, P., Bramanti, A., Pompa, P. P., Calabi, F., Cingolani, R., Rinaldi, R., Corni, S., Di Felice, R., Molinari, E., Verbeet, M. P., & Canters, G. W. (2005). Towards protein field-effect transistors: report and model of a prototype. *Adv. Mater., 17*, 816–822.

Mentovich, E. D., Belgorodsky, B., Kalifa, I., Cohen, H., & Richter, S. (2009). Large-scale fabrication of 4-nm-channel vertical protein-based ambipolar transistors. *Nano Lett., 9*, 1296–1300.

Mentovich, E. D., Belgorodsky, B., & Richter, S. (2011). Resolving the mystery of the elusive peak: negative differential resistance in redox proteins. *J. Phys. Chem. Lett., 2*, 1125–1128.

Mentovich, E., Belgorodsky, B., Gozin, M., Richter, S., & Cohen, H. (2012). Doped biomolecules in miniaturized electric junctions. *J. Am. Chem. Soc., 134*, 8468–8473.

Misra, N., Martinez, J. A., Huang, S.-C. J., Wang, Y., Stroeve, P., Grigoropoulos, C. P., & Noy, A. (2009). Bioelectronic silicon nanowire devices using functional membrane proteins. *Proc. Natl. Acad. Sci. USA, 106*, 13780–13784.

Park, H., Afzali, A., Han, S.-J., Tulevski, G. S., Franklin, A. D., Tersoff, J., Hannon, J. B., & Haensch, W. (2012). High-density integration of carbon nanotubes via chemical self-assembly. *Nat. Nanotechnol., 7*, 787–791.

Rinaldi, R., Biasco, A., Maruccio, G., Cingolani, R., Alliata, D., Andolfi, L., Facci, P., De Rienzo, F., Di Felice, R., & Molinari, E. (2002). Solid-state molecular rectifier based on self-organized metalloproteins. *Adv. Mater., 14*, 1449–1453.

Rinaldi, R., Biasco, A., Maruccio, G., Cingolani, R., Alliata, D., Andolfi, L., Facci, P., De Rienzo, F., Di Felice, R., Molinari, E., Verbeet, M., & Canters, G. (2003). Electronic rectification in protein devices. *Appl. Phys. Lett., 82*, 472–474.

Ron, I., Sepunaru, L., Itzhakov, S., Belenkova, T., Friedman, N., Pecht, I., Sheves, M., & Cahen, D. (2010). Proteins as electronic materials: electron transport through solid-state protein monolayer junctions. *J. Am. Chem. Soc., 132*, 4131–4140.

Tanford, C. (2004). *Ben Franklin stilled the waves*. Oxford: Oxford University Press.

Ulman, A. (2001). Formation and structure of self-assembled monolayers. *Chem. Rev., 96*, 1533–1554.

Willner, I., & Katz, E. (2000). Integration of layered redox proteins and conductive supports for bioelectronic applications. *Angew. Chem. Int. Ed., 39*, 1180–1218.

Zhao, J. W., Davis, J. J., Sansom, M. S. P., & Hung, A. (2004). Exploring the electronic and mechanical properties of protein using conducting atomic force microscopy. *J. Am. Chem. Soc., 126*, 5601–5069.

CHAPTER

Useful Notions in Electrochemistry

2

CHAPTER OUTLINE

2.0 Charged surfaces in water

When considering the action of electric fields on biological matter, one is unavoidably led to deal with electrochemistry. That is because, as thoroughly discussed in Chapter 1, biological molecules in particular and living beings in general are totally or partially located in water-based, salty environments. It is worth recalling that the average water content in living beings amounts to 70% (w/w), a figure that can even reach 90% in cell nuclei, and that "physiological solution", i.e., sea water, features 150 mM (NaCl equivalent) ionic strength.

Therefore, any electric field interacting with a biological object in its native environment will interact also with the ionic environment that hosts the object.

Biomolecular Electronics. http://dx.doi.org/10.1016/B978-1-4557-3142-8.00002-9

This fact is particularly important in relation to the goal of this book, which is to set the scene for direct electrical control of biological reactions and phenomena. Indeed, it is conceivable that such a possibility will be achieved by using electrically polarizable metal surfaces (electrodes) in contact with the physiological environment in which the biological reaction/phenomenon to be controlled takes place.

Electrically charged surfaces in physiological environments can be of both biological and exogenous nature (electrodes). Among the first, one can list the surfaces of complex biological macromolecules (e.g. proteins, nucleic acids), whose exposed surface is usually composed of chemical groups that can be electrically charged in physiological conditions (either positively, e.g., primary amines, or negatively, e.g., carboxyl acids, phosphate groups, etc.) and biological membranes, which can bear charges in the polar headgroups of their constituent (phospho)lipids, as well as in the exposed regions of membrane proteins.

The case of electrodes immersed in physiological fluids, which is pertinent to the use of direct electrochemistry for controlling the conformational or redox state of molecular components inside artificial biomolecular aggregates or living beings, will be the main subject of the present chapter.

We will start by examining the behavior of electrically charged surfaces in water with the goal of achieving a quantitative description of the electric potential and counterion concentration profile away from the surface of an electrically polarizable metal electrode in a physiological fluid and physiological-like solution.

To reach such a goal it is worth starting from a simpler situation that describes the effect of a charged surface immersed in pure water, i.e., without any added ions. In that case, only "counterions" will be present, that is, those ions possibly dissolved from the surface as it comes in contact with water (apart from H_3O^+ and OH^- ions coming from water dissociation).

The first step towards the stated goal is to establish some fundamental relations that describe the counterion distribution between two electrically charged surfaces in solution. In the following sections, we will then build on those first results to achieve a complete description of the more complex case, although more realistic and interesting for us, of an electrically polarizable surface immersed in a solution where added electrolytes are dissolved, such as the case of any biological buffer or fluid.

2.1 The Poisson-Boltzmann equation

To achieve the goal of a quantitative description of the potential and counterion density profile away from a charged surface, we first consider the chemical potential μ of any ion as:

$$\mu = ze\psi + k_B T \ln \rho, \tag{2.1}$$

where ψ represents the electrostatic potential and ρ the density of ions of valency z at any location x between two facing surfaces; Figure 2.1. We can put $\psi_0 = 0$ for $x = 0$ (midplane), and $\rho = \rho_0$ and $(d\psi/dx)_0 = 0$ for symmetry reasons.

At equilibrium, the chemical potential must be the same at any x, thus one obtains from equation (2.1) the Boltzmann distribution of counterions at any x, called the Nernst equation:

$$\rho = \rho_0 e^{-ze\psi/k_B T}. \tag{2.2}$$

We now need an additional important relation, which is the Poisson equation, giving the net excess charge density in x:

$$ze\rho = -\varepsilon_0 \varepsilon \left(d^2\psi/dx^2\right). \tag{2.3}$$

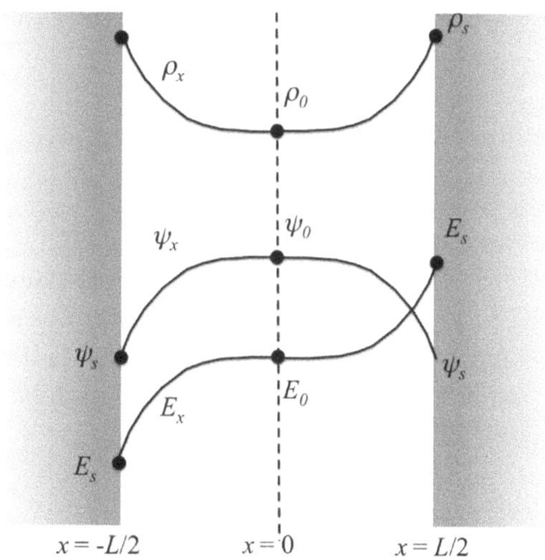

FIGURE 2.1

Scheme of two equally charged surfaces spaced by L and containing only counterions in the solution gap between them. The ion density ρ, the electric potential ψ, and the electric field E qualitative profiles are reported. Contact values are labeled with the subscript s.

Combining (2.2) and (2.3), we obtain the so-called Poisson-Boltzmann equation:

$$d^2\psi/dx^2 = -ze\rho/\varepsilon_0\varepsilon = -(ze\rho_0/\varepsilon_0\varepsilon)e^{-ze\psi/k_BT}. \tag{2.4}$$

This important equation gives the values of the potential ψ, the electric field E, and ρ at any x between the two surfaces.

Being a non-linear second-order differential equation, equation (2.4) needs two boundary conditions to be solved for ψ. Two suitable conditions come from (i) the symmetry consideration that the electric field must vanish at the midplane, and (ii) the requirement of overall electroneutrality, i.e., the total charge between the two surfaces has to equal that at the surfaces. That is why it is useful to determine first the values of potential, electric field and counterion concentration at the surfaces themselves, the so-called "contact values", ψ_s, E_s and ρ_s.

If σ is the surface charge density on each of the two surfaces separated by a distance L, the electroneutrality condition requires that:

$$\sigma = -\int_0^{L/2} ze\rho dx = +\varepsilon_0\varepsilon \int_0^{L/2} (d^2\psi/dx^2)^2 dx = -\varepsilon_0\varepsilon(d\psi/dx)_{L/2}$$
$$= +\varepsilon_0\varepsilon(d\psi/dx)_s = -\varepsilon_0\varepsilon E_s$$

which gives:

$$E_s = -\sigma/\varepsilon_0\varepsilon. \tag{2.5}$$

This result provides an important boundary condition that states that the contact value of the electric field is fully determined by the surface charge density, independently of the value of surface separation L.

Considering the ionic concentration, there exists an important relationship between its contact value and the value at the midplane ($L = 0$).

If we differentiate (2.2) and use (2.4) we obtain:

$$\frac{d\rho}{dx} = -\frac{ze\rho_0}{k_BT}e^{-ze\psi/k_BT}\left(\frac{d\psi}{dx}\right) = \frac{\varepsilon_0\varepsilon}{k_BT}\left(\frac{d\psi}{dx}\right)\left(\frac{d^2\psi}{dx^2}\right) = \frac{\varepsilon_0\varepsilon}{2k_BT}\frac{d}{dx}\left(\frac{d\psi}{dx}\right)^2, \tag{2.6}$$

therefore:

$$\rho_x - \rho_0 = \int_0^x d\rho = \frac{\varepsilon_0\varepsilon}{2k_BT}\int_0^x d\left(\frac{d\psi}{dx}\right)^2 = \frac{\varepsilon_0\varepsilon}{2k_BT}\left(\frac{d\psi}{dx}\right)_x^2$$

so that:

$$\rho_x = \rho_0 + \frac{\varepsilon_0 \varepsilon}{2k_B T}\left(\frac{d\psi}{dx}\right)_x^2, \tag{2.7}$$

which gives ρ at any x in terms of ρ_0 at the midplane and $(d\psi/dx)^2$ at x. At either surface, using (2.5) in (2.7) one obtains the contact value for ρ as:

$$\rho_s = \rho_0 + \sigma^2/2\varepsilon_0 \varepsilon k_B T. \tag{2.8}$$

This result shows that the contact value of the counterion density depends only on surface charge density σ and on counterion concentration at the midplane ρ_0. Particularly, it can never go below $\sigma^2/2\varepsilon_0\varepsilon k_B T$. For example, for an isolated surface in water, $\rho_0 \to 0$ and if we assume $\sigma = -0.2\ \mathrm{C\,m^{-2}}$ (1 e^- per 0.8 nm^2) at 293 K, one gets $\rho_S = \sigma^2/2\varepsilon_0\varepsilon k_B T = 7.0 \cdot 10^{27} m^{-3}$ which corresponds roughly to 12 M.

It is interesting to note that if one assumes that these counterions are located in a layer of thickness $\cong$ 0.2 nm, the previous figure corresponds to a counterion surface density of about one charge per 0.7 nm^2, a value very close to that assumed for σ. This fact means that most counterions that balance surface charges are always (i.e., irrespectively of potential profile) located within the first few ångstroms from the surface (Jönsson et al., 1980), giving rise to the well-known "electrical double layer" (Bard & Faulkner, 2001).

We are now in a position to solve the Poisson-Boltzmann equation, in the case of the presence of only counterions. The general case involving added ions will be faced later as a generalization of this case. Equation (2.4) can be satisfied by the following expression:

$$\psi = (k_B T/ze)\ln\left(\cos^2 Kx\right), \tag{2.9}$$

that is:

$$e^{-ze\psi/k_B T} = 1/\cos^2 Kx, \tag{2.10}$$

where K is a constant given by:

$$K = ze\left(\rho_0/2\varepsilon_0\varepsilon k_B T\right)^{1/2}. \tag{2.11}$$

We can note that the chosen tentative solution satisfies the symmetry requirements that $\psi = 0$ and $d\psi/dx = 0$ at $x = 0$ (midplane) for any value assumed by K.

If we now differentiate (2.9) and then use (2.5), we obtain for the electric field at any x and at the surfaces, respectively:

$$E_x = -d\psi/dx = (2k_B TK/ze)\tan kx, \tag{2.12}$$

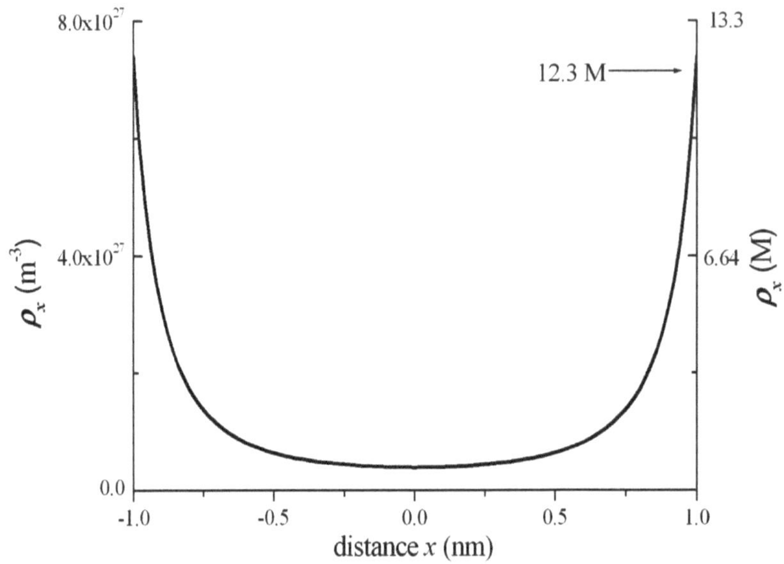

FIGURE 2.2

Counterion density distribution profile between two charged surfaces with $\sigma = 0.2$ C m^{-2} and spaced by $L = 2$ nm. The plot is obtained solving the Poisson-Boltzmann equation as given by equation (2.14).

and

$$E_S = -(d\psi/dx)_S = (2k_BTK/ze)\tan(KL/2) = -\sigma/\varepsilon_0\varepsilon. \tag{2.13}$$

Once K is determined from equations (2.12) and (2.13) as a function of σ and L, the counterion distribution profile is given by:

$$\rho_x = \rho_0 e^{-ze\psi/k_BT} = \rho_0/\cos^2 Kx. \tag{2.14}$$

Figure 2.2 shows the trend for the counterion density ρ_x at any position x between two facing surfaces, spaced by $L = 2$ nm, expressed in m^{-3} and M, on the left and right vertical axes, respectively. The curve is obtained by solving the Poisson-Boltzmann equation (2.14).

2.2 Charged surfaces in electrolytic solutions

We are now in a position to consider the case, by far more realistic, of a charged surface in contact with a solution containing added dissociated ions, as is the case of biological fluids.

Consider an isolated surface in an aqueous electrolyte solution. Put $x = 0$ at the surface. All the equations that we have derived in the case of two facing surfaces in a solution containing counterions only are still valid even in the case of different ions i, each of valence z_i, provided that one takes this fact explicitly into account by expressing the net charge density at any point x as $\sum z_i e \rho_{xi}$ and the overall ionic concentration as $\sum \rho_{xi}$. So equation (2.2), that is, the Boltzmann distribution of ions i at position x, becomes:

$$\rho_{xi} = \rho_{\infty i} e^{-z_i e \psi_x / k_B T} \tag{2.15}$$

which, at $x = 0$, relates the contact values of ρ and ψ as follows:

$$\rho_{0i} = \rho_{\infty i} e^{-z_i e \psi_0 / k_B T} \tag{2.16}$$

where $\rho_{\infty i}$ is the bulk concentration for ions i (at $x = \infty$), where $\psi_\infty = 0$.

Now, we can determine the total concentration of ions at an isolated surface with charge density σ. From equation (2.8) we can immediately write:

$$\sum_i \rho_{0i} = \sum_i \rho_{\infty i} + \sigma^2 / 2\varepsilon_0 \varepsilon k_B T. \tag{2.17}$$

Thus, for a surface with a charge density $\sigma = -0.2$ Cm^{-2} at 25 °C, the term $\sigma^2 / 2\varepsilon_0 \varepsilon k_B T = 7 \cdot 10^{27} m^{-3} = 11.64$ M. In the case of a 1:1 electrolyte such as KCl, whose bulk molar concentration is [KCl], the surface concentration of ions is:

$$\begin{aligned} [\mathrm{K}^+]_0 + [\mathrm{Cl}^-]_0 &= 11.64 + [\mathrm{K}^+]_\infty + [\mathrm{Cl}^-]_\infty = 11.64 + 2[\mathrm{K}^+]_\infty \\ &= 11.64 + 2[\mathrm{KCl}]\ \mathrm{M}, \end{aligned}$$

whereas, in the case of a 2:1 electrolyte such as $CaCl_2$ with a bulk molar concentration $[CaCl_2]$:

$$\begin{aligned} [\mathrm{Ca}^{2+}]_0 + [\mathrm{Cl}^-]_0 &= 11.64 + [\mathrm{Ca}^{2+}]_\infty + [\mathrm{Cl}^-]_\infty = 11.64 + 3[\mathrm{Ca}^{2+}]_\infty \\ &= 11.64 + 3[\mathrm{CaCl}_2]\ \mathrm{M}. \end{aligned}$$

Now, we are in a position to find the relation between the surface charge density σ and the surface potential ψ_0. Substituting equation (2.15) in (2.16), one obtains, for the case of a mixed KCl + $CaCl_2$ electrolyte:

$$\begin{aligned} \sigma^2 &= 2\varepsilon_0 \varepsilon k_B T \left(\sum_i \rho_{0i} - \sum_i \rho_{\infty i} \right) = \\ &= 2\varepsilon_0 \varepsilon k_B T \Big\{ [K^+]_\infty e^{-e\psi_0/k_B T} + [Ca^{2+}]_\infty e^{-2e\psi_0/k_B T} \\ &\quad + [Cl^-]_\infty e^{+e\psi_0/k_B T} - [K^+]_\infty - [Ca^{2+}]_\infty - [Cl^-]_\infty \Big\}. \end{aligned}$$

Now, noting that $[Cl^-]_\infty = [K^+]_\infty + 2[Ca^{2+}]_\infty$, the preceding becomes:

$$\sigma^2 = 2\varepsilon_0 \varepsilon k_B T \Big\{ \left[K^+\right]_\infty \left(e^{-e\psi_0/k_BT} + e^{+e\psi_0/k_BT} - 2\right) + \left[Ca^{2+}\right]_\infty \left(e^{-2e\psi_0/k_BT} + 2e^{+e\psi_0/k_BT} - 3\right)\Big\},$$

thus, finally, one obtains the Grahame equation:

$$\sigma = \sqrt{8\varepsilon_0 \varepsilon k_B T} \sinh(e\psi_0/2k_BT) \times \left\{ \left[K^+\right]_\infty + \left[Ca^{2+}\right]_\infty \left(2 + e^{-e\psi_0/k_BT}\right)\right\}^{1/2} Cm^{-2}. \quad (2.18)$$

The Grahame equation (Grahame, 1953) allows us to calculate σ if ψ_0 is known, or vice versa, from which the individual counterion surface concentration can be determined using equation (2.16). It is worthy of note that this equation just relates σ and ψ_0, it does not predict either of them, the other being known.

In case of low potentials, the Grahame equation simplifies to:

$$\sigma = \varepsilon_0 \varepsilon \kappa \psi_0, \quad (2.19)$$

where

$$\kappa = \sqrt{\sum_i \rho_{\infty i} e^2 z_i^2 / \varepsilon_0 \varepsilon k_B T}. \quad (2.20)$$

This result implies that the potential becomes proportional to the surface charge density. Equation (2.19) corresponds to that for a capacitor whose plates are separated by a distance $1/\kappa$, with charge densities of $\pm\,\sigma$, and potential difference ψ_0.

This analogy with a charged capacitor has given rise to the expression “diffuse electric double layer” to mean the ionic atmosphere near a charged surface. Its characteristic thickness, or decay length, is called the “Debye length” and corresponds to $1/\kappa$. This parameter is very important and useful to quantify the extent of a potential away from a charged surface, a concept that we will use on several occasions and that will be crucial in some applications concerned with affecting the conformation of biomolecules lying at a finite distance off a polarized metal electrode surface, as in the case of molecular layers assembled at an electrode’s surface.

Its magnitude depends only on the properties of the solution and not on properties such as the potential or charge density of the surface.

For instance, in the case of a KCl solution, $1/\kappa = 30.4$ nm at 10^{-4} M, 9.6 nm at 1 mM, 0.96 nm at 10^{-2} M and 0.3 at 1 M. Thus, by adjusting the electrolyte concentration (in a range spanning four orders of magnitude) it is possible to vary the Debye length by two orders of magnitude.

2.3 Potential and ion concentration away from a charged surface

The potential gradient at any x away from a charged surface is given by equation (2.7):

$$\sum_i \rho_{xi} = \sum_i \rho_{\infty i} + \frac{\varepsilon_0 \varepsilon}{2k_B T}\left(\frac{d\psi}{dx}\right)_x^2 \tag{2.21}$$

In the case of a 1:1 electrolyte, it gives: $d\psi/dx = \sqrt{8k_BT\rho_{\infty i}/\varepsilon_0\varepsilon\, sinh(e\psi_x/2k_BT)}$, which may be integrated using the integral $\int coshxdx = \ln(tanh(x/2))$, to give:

$$\psi_x = \frac{2k_BT}{e}\ln\left(\frac{1+\gamma e^{-\kappa x}}{1-\gamma e^{-\kappa x}}\right) \approx \frac{4k_BT}{e}\gamma e^{-\kappa x}, \tag{2.22}$$

where $\gamma = \tanh(e\psi_0/4k_BT)$.

This is known as the Gouy-Chapman theory. At high potential $\gamma \rightarrow 1$, while it reduces to the so-called Debye-Hückel equation, $\psi \approx \psi_0 e^{-\kappa x}$, for potentials less then 25 mV (Verwey & Overbeek, 1948; Hiemenz & Rajagopalan, 1997). Again, the Debye length plays the role of a characteristic decay length of the potential.

We have derived the preceding equations under the assumption of a symmetrical, 1:1 electrolyte solution such as KCl or NaCl. The corresponding equations for asymmetrical electrolytes, such as 2:1 (e.g. $CaCl_2$) or 1:2 (Na_2SO_4), are more complicated than equation (2.22) but reduce to $\psi \approx \psi_0 e^{-\kappa x}$ for small values of ψ_0.

We now have all the equations needed to compute the ionic distribution away from an isolated charged surface. In the case of 1:1 electrolyte, this is obtained by inserting equation (2.22) into (2.15). Figure 2.3 shows the trend of ψ_x and ρ_x for a 0.15 M 1:1 electrolyte. Note that for such a high surface charge density and potential, the counterion density approaches the bulk value much faster than would be suggested by the value of the Debye length. Indeed, in these conditions, the counterion density near the surface is largely independent of the bulk electrolyte concentration.

The Gouy-Chapman or Debye-Hückel theories and their implications concerning the potential and counterion profiles off a charged surface represent for us solid ground on which we are going to build most of the results

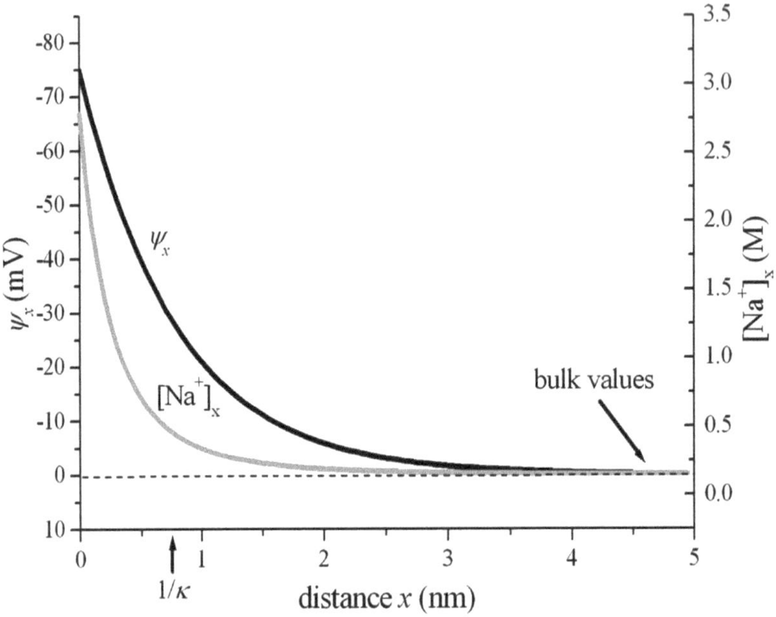

FIGURE 2.3

Potential ψ_x and counterion density $[Na^+]_x$ profiles away from a surface with surface charge density $\sigma = -0.0922$ C m^{-2}, $\psi_0 = -75$ mV, as resulting from the Grahame equation (2.18) and the Gouy-Chapman model (2.22). Note that counterion concentration decreases much faster than potential away from the surface.

and perspective approaches that will be shown in this book. In fact, the predictions of those theories set the stage for affecting the redox state of both freely diffusing and surface immobilized (bio)molecules and enable the design of experiments aimed at manipulating the conformation of biomolecules that can respond to an externally applied electric field. A crucial aspect in the success of such approaches will be concerned with the proper design of aqueous electrolytic solutions in order to enable a suitable spatial extent of an electric field arising from an electrode surface. In terms of the concepts we have introduced here, it will be necessary to adjust the ionic strength of the working solution to the proper value to tune the Debye length to the desired extent. This requirement, unfortunately, cannot be trivially achieved just by lowering the ionic strength to a minimum (the Debye length in pure water at pH 7 is 960 nm) since another requirement has to be taken into account, connected to the resistance of the electrolytic working solution. Indeed, it cannot decrease below a given threshold if the potential of the metal electrode we are using is to be controlled by an external voltage source with the required accuracy and speed.

All of these considerations call for further insight into the physical chemistry of electrical polarizable surfaces in electrolytic solutions, which will be the subject of the following sections.

2.4 Reactions at electrodes

We now start considering phenomena that can take place when we immerse an electrode in an electrolytic solution and apply an external potential to it with respect to another (reference) electrode also immersed in the solution.

Two basic kinds of processes can take place at an electrode exposed to an electrolytic solution. One of them comprises reactions in which charges are transferred across the interface between metal and solution. If we consider electrons, their transfer causes the occurrence of oxidation or reduction phenomena. In this first type of reaction, where a current flows across an electrochemical interface, the amount of reaction products is proportional to the amount of flowing charges. Thus, these are called *faradaic* processes, since they obey Faraday's law. Electrodes at which faradaic processes take place are also referred to as charge-transfer electrodes.

When an electrode exchanges electrons with a molecular species, either freely diffusing in solution or bound to the electrode's surface (e.g., in the case of a self-assembled monolayer of redox molecules), a heterogeneous redox reaction takes place.

The electron exchange takes place from around the Fermi level of the (metal) electrode to the lowest unoccupied molecular orbital (LUMO) of a redox species or from the highest occupied molecular orbital (HOMO) to around the Fermi level of the electrode. In the first case we are dealing with a reduction current, and in the second one with an oxidation current. Both these situations can be matched by adjusting the electrode potential (electron energy level) to the proper value (around the standard potentials E_0 for reduction or for oxidation) by means of an external voltage source connecting the electrode to another (reference) electrode in solution (see section 2.5). Figure 2.4 shows schematically the process of both heterogeneous oxidation and reduction.

Generally speaking, there are potential ranges in which charges can cross the electrochemical barrier, giving rise to a current. Out of those potential windows, charge transfer reactions do not occur since they are

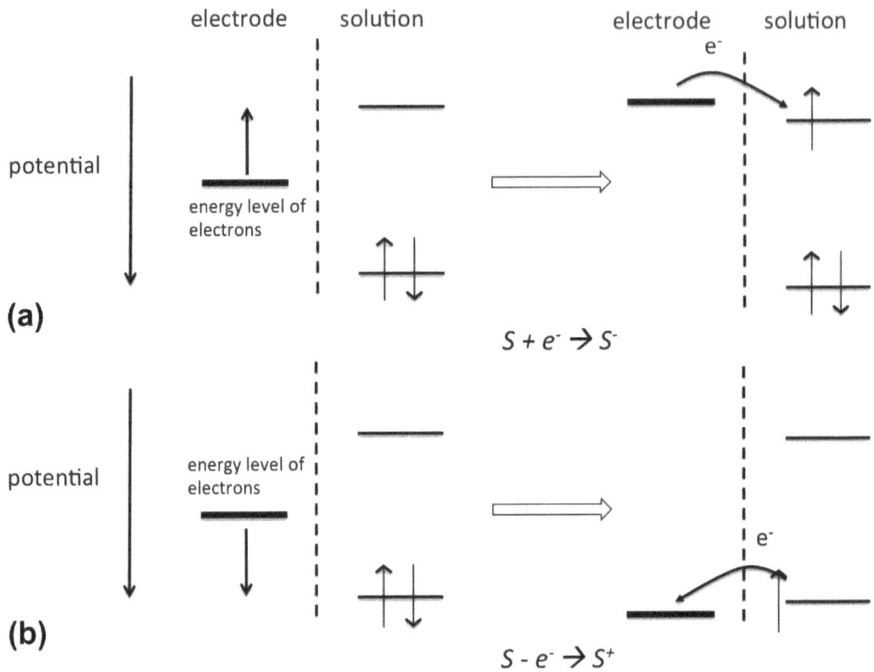

FIGURE 2.4

Scheme of heterogeneous reduction (a) and oxidation (b) of a species *S* giving rise to faradaic currents.

kinetically or thermodynamically unfavorable. In the described conditions, processes such as adsorption and desorption can still occur and can alter the structure of the interface as a function of the applied potential or solution composition. These kinds of phenomena are called *non-faradaic* and, even if no electrons cross the interface, can give rise to sizable "charging" or "uncharging" currents, capacitive in nature, and stemming from the electrical double layer that has been introduced in 2.2. The corresponding "effective" capacitor that is associated with the double layer can acquire specific capacitance values in the range 10–40 $\mu F/cm^2$ (Bard & Faulkner, 2001). At variance with an electric capacitor, however, these values are often a function of the potential applied across the electrochemical interface.

Of course, faradaic and non-faradaic currents can appear together when electrode reactions take place. Both kinds of reactions will be used in the following chapters to condition the state of a molecule or of a molecular system as well as to retrieve critical information about the formal potential of redox couples, diffusing or bound to a surface, to measure the electron-transfer rate constant of specific reactions, or to find out details on the molecular structure of adsorbed layers. They will be used also to switch the conformation of biological macromolecules with the help of the electric field associated with an electrically polarized electrode.

Whereas it is beyond the scope of this book to present an exhaustive treatment of electrode reaction kinetics, especially based on macroscopic concepts, for which we address the readers to any of the authoritative treatises reviewing the field (see e.g., Bard & Faulkner, 2001), it is instead our intention to review briefly the Marcus microscopic approach to electron-transfer reactions before introducing some of the electrochemical techniques and tools that turn out to be particularly useful when dealing with redox molecules or with molecules that can interact with electric fields.

2.4.1 The Marcus theory of electron transfer

Before facing issues related to the experimental use of electrochemistry to characterize and affect electron-transfer reactions by direct electrochemistry, we find it useful to provide, here, a synthetic account of the ideas developed by Marcus (Marcus, 1956, 1968) in an effort to provide a microscopic description of how molecular structure and environment can affect the electron-transfer processes. This very important theory represents the basis for the interpretation of most of the macroscopic and single-molecule electrochemical experiments.

We will refer here to the concepts of inner-sphere and outer-sphere electron-transfer reactions at electrodes. This terminology has been borrowed from that used in electron-transfer processes of coordination compounds where inner-sphere reactions occur in activated complexes where ions share a ligand. Outer-sphere reactions refer instead to a reaction in which the original coordination spheres (those of reactants) are preserved in the activated complexes. Similarly, an inner-sphere electrode reaction is a heterogeneous electron-transfer reaction where there is a strong interaction of reactants, intermediates or products with the electrode (accompanied by a specific adsorption of species on the electrode). In contrast, in an outer-sphere electrode reaction reactants or products do not interact strongly with the electrode surface, being at least one solvent layer apart from it. Due to their lower specificity, outer-sphere reactions can be described in a more general fashion than inner-sphere ones and that is the reason why the corresponding theories are much more developed.

If we consider an outer-sphere, single electron transfer from an electrode to species O, as a result a product R will form. Such a heterogeneous reaction has much in common with the homogeneous reaction of O to R by a reductant R′ that results in $O + R' \rightarrow R + O'$. It is useful to consider these two situations in the same theoretical context. Both homogeneous and heterogeneous electron-transfer reactions are radiationless electronic rearrangements of reacting species. As such, an electron has to move from an initial state to a final one of the same energy. This aspect is a key one in the treatment.

Another important characteristic, common to most microscopic theories of electron transfer, is the hypothesis that reactants and products do not change their configuration during the transfer of the electron. This description refers to the Franck-Condon principle, for which nuclear momenta and positions do not change on the time scale of electronic transitions. Therefore, species O and R have the same nuclear configuration at the moment of electron transfer.

Now let us consider a plot of the standard free energy of species O and R as a function of a reaction coordinate; Figure 2.5. The aim here is to obtain an expression for the standard free energy of activation $\Delta G^{\ddagger}$ as a function of the structural parameters of the reactants. In either the homogeneous or the heterogeneous case, let us consider a reaction taking place on a multidimensional surface defining the standard free energy of the system as a function of the nuclear coordinates (i.e., the relative atomic positions) of reactants, products and solvent.

Changes in these coordinates arise from rotations and vibrations in O and R species as well as from fluctuations in position and orientation of solvent molecules. We use a reaction coordinate q, focusing on the energetically favored pathway between reactants and products. We also assume that (i) the reactant O is fixed in some position with respect to the electrode (or that the reactants' mutual distance is fixed, if we consider a homogeneous biomolecular reaction) and (ii) that the standard free energy for O and R, G_0^O and G_0^R, depends quadratically on q (Marcus & Siddarth, 1992):

$$G_0^O(q) = (c/2)(q - q^O)^2 \tag{2.23}$$

$$G_0^R(q) = (c/2)(q - q^R)^2 + \Delta G_0 \tag{2.24}$$

where q^O and q^R are the equilibrium coordinates for the atomic configurations of O and R and c is a proportionality constant. ΔG_0 will be either the reaction free energy for a homogeneous reaction or $F(E - E_0)$ for a heterogeneous one.

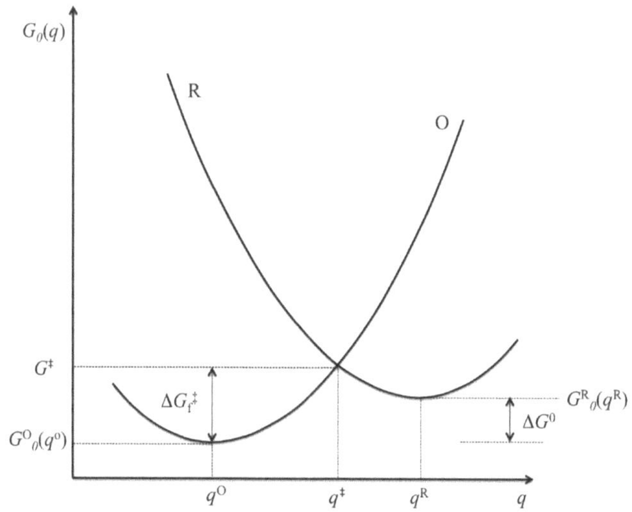

FIGURE 2.5

Standard free energy as a function of reaction coordinate for an electron-transfer reaction. The scheme is valid for both heterogeneous and homogeneous reactions. In the heterogeneous case, closer to the scope of the present book, the curve for O is the sum of the energies for species O and for an electron on the electrode at the Fermi level as determined by the potential E. Therefore, $\Delta G_0 = F(E - E_0)$.

With reference to Figure 2.6, the transition state is the position along the reaction coordinate where O and R have the same configuration, say $q^{\ddagger}$. According to the Franck-Condon principle, electron transfer takes place at this position.

At the transition state, the free energies are given by

$$G_0^O(q^{\ddagger}) = (c/2)(q^{\ddagger} - q^O)^2 \tag{2.25}$$

$$G_0^R(q^{\ddagger}) = (c/2)(q^{\ddagger} - q^R)^2 + \Delta G_0 \tag{2.26}$$

Since $G_0^O(q^{\ddagger}) = G_0^R(q^{\ddagger})$, equations (2.25) and (2.26) can be solved for $q^{\ddagger}$, giving

$$q^{\ddagger} = \frac{(q^R + q^o)}{2} + \frac{\Delta G_0}{c(q^R - q^o)} \tag{2.27}$$

The free energy of activation for reduction of the species O is

$$\Delta G_f^{\ddagger} = G_0^O(q^{\ddagger}) - G_0^O(q^O) = G_0^O(q^{\ddagger}) \tag{2.28}$$

since $G_0^O(q^O) = 0$ from its definition (2.23).

Inserting the expression (2.27) in equation (2.25) gives

$$\Delta G_{\mathrm{f}}^{\ddagger} = \frac{c\left(q^R - q^o\right)^2}{8} + \left[1 + \frac{2\Delta G_0}{c\left(q^R - q^o\right)^2}\right]^2 \tag{2.29}$$

If one defines $\lambda = (c/2)(q^R - q^O)^2$ it results in

$$\Delta G_f^{\ddagger} = \frac{\lambda}{4}\left(1 + \frac{\Delta G_0}{\lambda}\right)^2 \tag{2.30}$$

or, for a heterogeneous reaction

$$\Delta G_f^{\ddagger} = \frac{\lambda}{4}\left(1 + \frac{F(E - E_0)}{\lambda}\right)^2 \tag{2.31}$$

The parameter λ is the *reorganization energy*, which represents the energy needed to change the reactant and solvent nuclear configurations into those of the product state. Usually, it is split into inner λ_i (contribution from the reorganization of species O) and outer λ_o (contribution from solvent reorganization) components,

$$\lambda = \lambda_i + \lambda_o. \tag{2.32}$$

Under the assumption that the normal mode of the reactant remains harmonic over the distortion range on interest, it is in principle possible to calculate λ_i by summing over the normal vibrational modes of the reactant, obtaining,

$$\lambda_i = \sum_j \frac{1}{2} c_j \left(q_j^o - q_j^R\right)^2 \tag{2.33}$$

the c's being force constants and the q's displacements in normal coordinates.

As to λ_o, it is usually estimated under the assumption that the solvent can be modeled as a dielectric continuum and the reactant is a sphere of radius a^O.

In the case of a heterogeneous reaction it takes the form

$$\lambda_o = \frac{e^2}{8\pi\varepsilon_0}\left(\frac{1}{a^0} - \frac{1}{R}\right)\left(\frac{1}{\varepsilon_{op}} - \frac{1}{\varepsilon_s}\right) \tag{2.34}$$

where ε_{op} and ε_s are the optical and static dielectric constants, respectively, and R is twice the distance of the center of the molecule from the electrode.

Lastly, in the case of a homogeneous reaction,

$$\lambda_o = \frac{e^2}{4\pi\varepsilon_0}\left(\frac{1}{2a^I}+\frac{1}{2a^{II}}-\frac{1}{d}\right)\left(\frac{1}{\varepsilon_{op}}-\frac{1}{\varepsilon_s}\right) \tag{2.35}$$

where a^I and a^{II} are the radii of the reactants (O and R′) and $d = (a^I + a^{II})$. λ assumes typical values in the range 0.5−1 eV.

Both Marcus theory and most of the concepts introduced in this section will turn out to be essential in describing the behavior of the electrochemistry of redox biomolecules and cofactors and of their single-molecule characterization by ECSTM (see section 4.5.1).

2.5 Electrochemical tools

We are interested, basically, in tools and techniques used to measure an electrochemical current as a function of an applied potential and in controlling the potential of an electrode in an electrochemical cell. Irrespectively of the experimental technique, the first step to take is to introduce the way in which such a potential can be applied to an electrode in solution, that is, to introduce the concept of potentiostatic control of an electrode potential.

This concept is at the basis of any (potentiostatic) experimental approach and, from a physical standpoint, it can be regarded as an external source of free energy used to feed an electrode in order to drive its potential against a reference.

Any electrochemical measurement or conditioning of molecular systems that we will have to deal with will be performed in a standard three-electrode electrochemical cell.

Such a cell will be equipped with (i) a *working electrode* (WE), i.e., an electrode the potential of which we want to control, driving the reactions taking place at its surface; (ii) a *reference electrode* (RE), i.e., an electrode which approaches ideal polarizability (e.g., SCE, Ag/AgCl) and with respect to which the potential is controlled, and (iii) a *counter, auxiliary* or *sacrificial electrode* (CE) which passes the current needed to control the potential at the WE. The cell is characterized by a cell resistance R_s, and by the current flowing between CE and WE, i.

Figure 2.6 describes the potential drop in a three-electrode cell in the hypothesis of a linear potential drop (not always true).

Potentiostatic control of the WE is to be achieved by varying the current i flowing between the CE and the WE, i.e., the CE potential iR_s, in order to

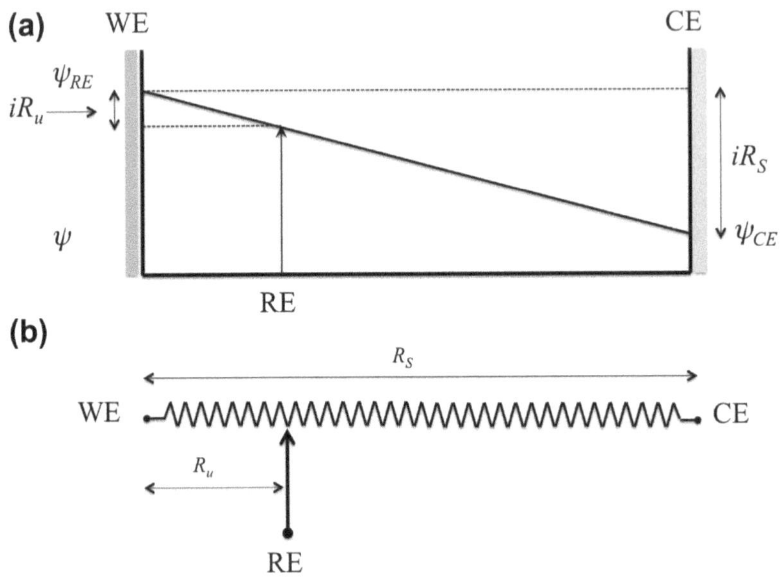

FIGURE 2.6

(a) Potential drop between working and counter electrodes in solution.
(b) Equivalent potentiometer.

maintain the potential drop between WE and RE at a desired value. The described set-up has a circuit analogy in a potentiometer.

2.5.1 Potentiostats

The external circuitry that is used to achieve the goal of driving the WE potential at the desired value is a potentiostat. There exist several possible designs for a potentiostat, among which the simplest is that based on an active network that is able to control the potential of a fixed point C in a network of impedances, even if those impedances fluctuate. Figure 2.7a shows the scheme of such a circuit. In this scheme, since the inverting input is at virtual ground, the point C will be at $-V_i$ vs ground. The operational amplifier will adjust its output in order to maintain this condition by varying the current through the impedances Z and R. The described function is exactly that performed by a potentiostat. Note the input potential source placed in the feedback circuit.

Particularly, since the current through Z must also pass through R, the total output voltage V_o, is $i_o(Z+R)$. Since $i_o = -V_i/R$, the output voltage will be:

$$V_o = -V_i\left(\frac{Z+R}{R}\right). \tag{2.36}$$

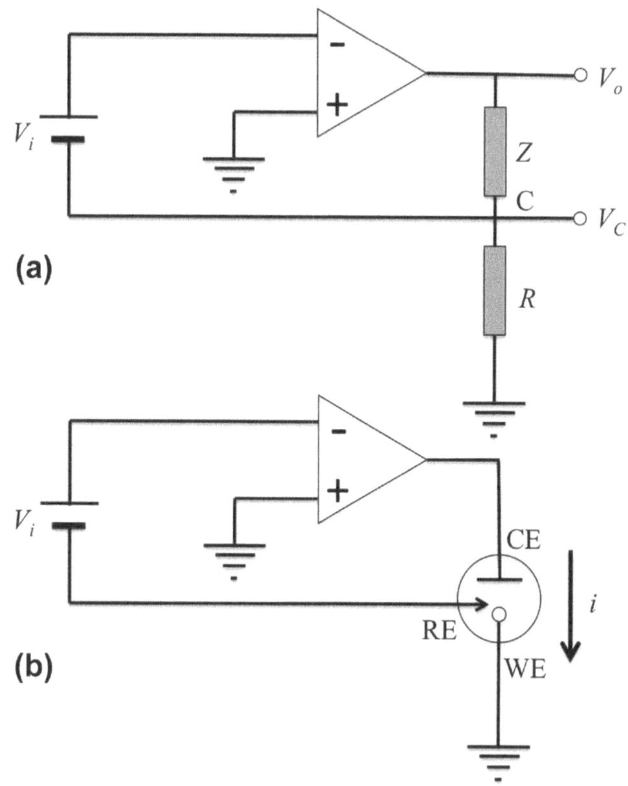

FIGURE 2.7

(a) Scheme of a circuit for controlling the potential at a point *C* in a network of impedances (*Z*, *R*) independently of their fluctuations. (b) A simple potentiostat based on the control circuit in (a).

Figure 2.7b shows the scheme of a simple potentiostat based on the control circuit of Figure 2.7a. Since RE is always at $-V_i$ vs ground, WE will be always be at V_i vs RE.

Although the described design can accomplish the task of controlling the potential of RE (or WE) also in the presence of a fluctuating network of impedances, it suffers some drawbacks. First, neither terminal of the potential source is a true ground. As a consequence, the function generator providing the signal for potential control would have to possess a differential floating output. Unfortunately, this is often not the case. Second, the controlling input waveforms are often complicated and should be synthesized ad hoc.

A special design called "adder potentiostat" solves both the aforementioned problems and, as a consequence, results in the most widely used potentiostat by far. Its basic scheme is given in Figure 2.8.

The sum of all the currents at the summing point S must equal zero,

$$-i_{RE} = i_1 + i_2 + i_3 \tag{2.37}$$

and, since S is a virtual ground, it results in:

$$-V_{RE} = V_1\left(\frac{R_{RE}}{R_1}\right) + V_2\left(\frac{R_{RE}}{R_2}\right) + V_3\left(\frac{R_{RE}}{R_3}\right). \tag{2.38}$$

As in the previous case of the simpler design, $-V_{RE}$ is the potential of WE vs RE. Therefore, the circuit maintains the WE at a potential given by the weighted sum of the inputs. If all input resistors are equal it results in:

$$V_{WE}(vs\ RE) = V_1 + V_2 + V_3. \tag{2.39}$$

Therefore, it is clear that the adder potentiostat fixes the two drawbacks mentioned above; here, inputs are given with respect to a true ground and the synthesis of arbitrary wavefunctions to implement various electrochemical measurements is now straightforward.

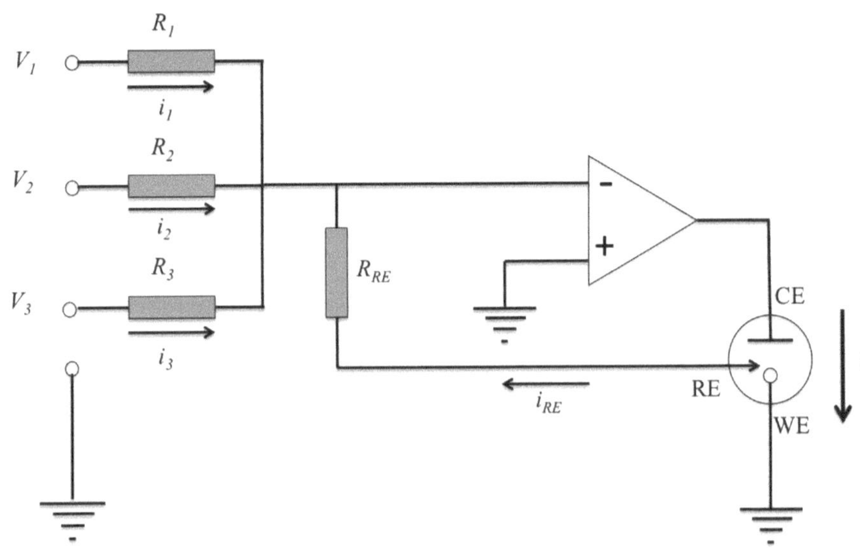

FIGURE 2.8

Scheme of a basic adder potentiostat.

2.5.2 Bipotentiostats

Some electrochemical experiments require the use of two working electrodes and the simultaneous control of their potential vs a single reference. A device that meets these requirements is called a bipotentiostat. We will often make reference to its functionality in Chapter 4, when we will discuss the set-up of and experiments performed with the electrochemical scanning tunneling microscope (ECSTM). Typical bipotentiostats take advantage of the adder potential principle for controlling one WE potential and then use it as a reference for driving the second WE potential. The use of other amplifiers as inverters and zero-shifting stages allows one to supply the desired potential V_2 at the input, independently of the value of V_I. With a bipotentiostat it is therefore possible to drive simultaneously different electrochemical reactions at two working electrodes.

2.5.3 Galvanostats

In electrochemical experiments where one is interested in the potential of the working electrode with respect to a reference electrode, a galvanostat is to be used. It allows control of the current flowing through a cell rather than the potential. This operation is, in general, easier than controlling a potential since only two elements of the cell, WE and CE, are involved in the control circuit.

Figure 2.9 shows the scheme of a simple galvanostat based on an inverter circuit. Here, cell impedance plays the role of a feedback resistance. The sum of the currents at S gives:

$$i_{cell} = -i_{in} = \frac{-V_i}{R}, \tag{2.40}$$

so that the cell current is controlled by V_i. The presence of the voltage follower F gives access to the working electrode potential. As in the case of the adder potentiostat, the input stage can accommodate more resistors, hence summing different currents that will converge to drive i_{cell}.

2.6 Electrochemical techniques

Potentiostats, bipotentiostats and galvanostats find ready application in the various techniques that are routinely used in electrochemical experimentation. The relevant experimental techniques are really numerous and it is beyond the scope of this book to review all of them and the physical

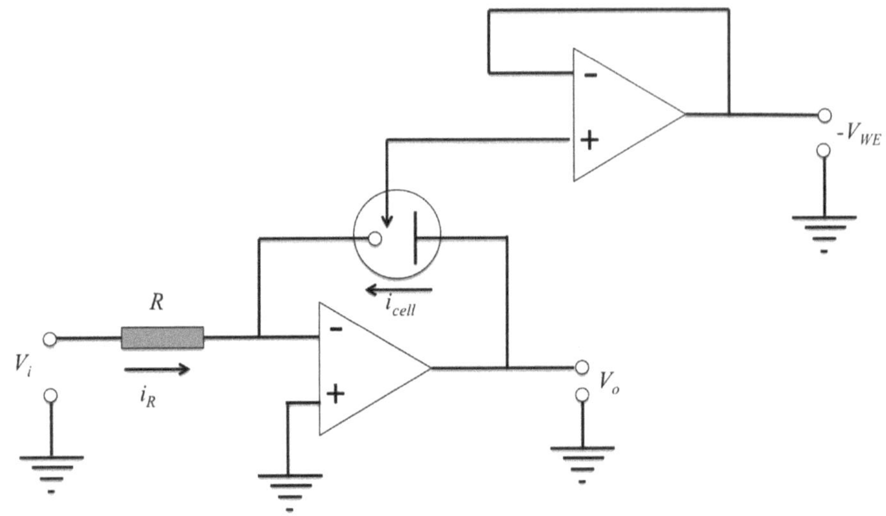

FIGURE 2.9

Scheme of a simple galvanostat based on an inverter circuit.

chemistry behind them. Again, there exist superb books and reviews that treat such a topic in due form and extent and the interested reader is referred to those for further reading (Bard & Faulkner, 2001; Hamann et al., 1998). In what follows we will introduce some of them of particular relevance for our work.

2.6.1 Potential sweep methods: LSV and CV

Our aim, here, is to focus on some of the most popular techniques that have been used quite extensively in the characterization of the molecular samples and assemblies included in the present book or that are proposed for some of the future developments of the subject of controlling biological reactions and phenomena by direct electrochemistry.

Among the techniques that we will take into consideration, a prominent place is occupied by voltammetric approaches. These are experimental techniques that are operated by sweeping the potential of WE vs RE, while recording the corresponding current that flows between WE and CE.

In order to adhere to a widespread convention in electrochemistry, we will use the symbol E in place of V to denote potential.

A complete electrochemical characterization of a system can be achieved by stepping the potential while measuring $i = i(t)$ at each potential step. This approach provides an i-t-E surface; however, it is generally very time-consuming, cumbersome and not very effective because of the large amount of data needed for a complete characterization. Indeed, whereas it is not easy to recognize the presence of different waves from $i = i(t)$ curves, to obtain well-resolved $i = i(E)$ curves it is often necessary to step the potential each mV. A more effective and information-rich approach is that of sweeping E in time while recording $i = i(E)$ curves, directly.

Usually, the potential is swept linearly with time, with typical sweep rates ranging from 10 to 1000 V/s in the case of conventional electrodes, and up to 10^6 V/s when using ultramicro-electrodes (UME). In practice, what it is done is to measure current vs potential curves, which corresponds to recording current vs time. By doing this, one performs linear potential sweep chronoamperometry or, more simply, linear sweep voltammetry (LSV).

An LSV experiment is illustrated in Figure 2.10. Let us consider a diffusing redox species in its oxidized form. If we start the experiment at a potential well positive of E_0' for the reduction, Figure 2.10b, sweeping the potential as reported in Figure 2.10a, only nonfaradaic currents will contribute to the overall current flow for some time. Once the electrode potential reaches the proximity of E_0', the reduction starts and faradaic current begins to flow. As the potential continues to go more negative, the surface concentration of the oxidized species has to fall. This causes an increase of the flow of oxidized species from solution bulk, hence of the current. Further decreasing the potential to E_0', the surface concentration of the oxidized species becomes almost zero and the mass transfer rate from solution increases, reaching a maximum, along with the current; then it declines due to depletion. The described mechanism shows itself as a peak in the $i = i(E)$ curve.

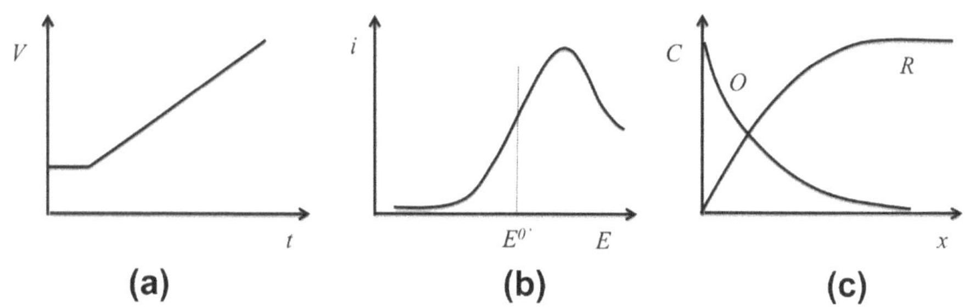

FIGURE 2.10

(a) Potential ramp as a function of time; (b) LSV wave; (c) concentration profiles for oxidized (O) and reduced (R) species as a function of the distance x from the WE surface.

Concentration profiles close to WE are similar to those reported in Figure 2.10c. Let us consider what happens if we reverse the potential scan. The potential now sweeps in a positive direction and near the surface there is a large concentration of the oxidizable species. As the potential grows positive, there will be more and more reduced species in the vicinity of WE; a situation similar to that described for the forward sweep will take place and a cathodic wave will develop for analogous reasons. We have performed cyclic voltammetry (CV), an experiment that corresponds to two LSV measurements in two different sweep directions. Figure 2.11 shows it in detail.

CV is a very popular electrochemical characterization technique that allows one to retrieve a number of pieces of information about diverse redox systems and parameters.

It is useful to bear in mind that for a reversible voltammetric wave (i.e., that arising from voltammetry of Nernstian systems; Bard & Faulkner, 2001) the peak potential, i.e. the potential at which the peak is located, is independent of the scan rate. The current peak intensity, furthermore, is proportional to the square root of the scan rate, indicating a control of the electrochemical reaction by diffusion.

In the next section we will focus on useful approaches developed by Laviron aimed at retrieving information on the adsorption state of a redox species as well as on its heterogeneous electron-transfer rate constant. Both these kinds of information give extremely useful initial hints about unknown redox systems such as redox metalloproteins or adsorbed redox cofactors.

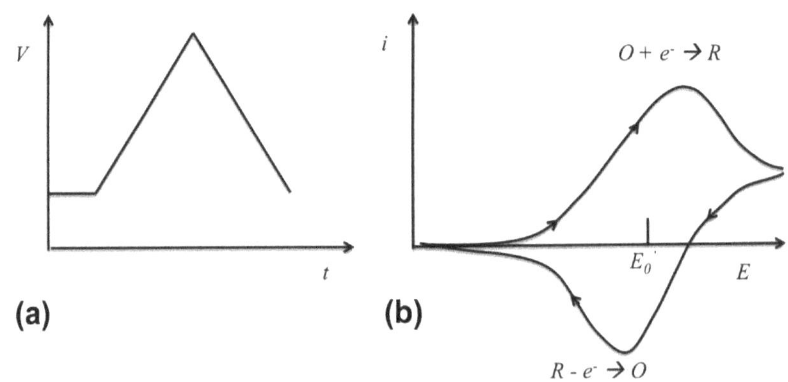

FIGURE 2.11

(a) Sweeping of potential vs time in cyclic voltammetry; (b) cathodic and anodic waves in cyclic voltammetry.

2.6.2 Laviron's formalism

In a seminal work dating back to the late seventies, E. Laviron developed a theory for interpreting and extracting information from voltammetric measurements performed on redox systems that are strongly adsorbed on a WE (Laviron, 1979). That is the case of self-assembled monolayers of molecules bearing a redox moiety, as well as (sub)monolayers of redox metalloproteins chemisorbed or physisorbed on an electrode surface. These kinds of systems are of central interest to the ideas developed in this book, so it is worthwhile recalling the main results obtained by Laviron that apply to them.

The theory at issue deals with a class of redox systems subjected to heterogeneous oxidation and reduction in the hypothesis that diffusion does not play any role, i.e., mass-transfer kinetics is not at play. Under those conditions and assuming an adsorption isotherm such as the Langmuir or linearized Langmuir one, one finds that the peak current i_p is related to the scan speed v (i.e., the potential sweep speed) in a linear fashion, at variance with what is known for diffusing reversible systems, where the aforementioned hypotheses are not true and the peak current is proportional to $v^{1/2}$. Therefore, this result provides a method for a ready proof of the adsorption state of a redox species and is often used as an early characterization of an adsorbed redox layer. In practice, measuring CVs as a function of the scanning speed and then plotting i_p as a function of v provides the answer.

Figure 2.12a illustrates the trend of both cathodic (i_{pC}) and anodic (i_{pA}) as a function of the scan rate for a molecular system of immobilized hydroquinone molecules.

Furthermore, under the same assumptions, Laviron has demonstrated (Laviron, 1979) that it is possible to extract the electron-transfer rate constant k and the transfer coefficient α (or the number of electrons n involved in the reaction if α is assumed as known) by plotting the position of the cathodic and anodic peak potential as a function of the ln$v$. For $|\Delta E_p| > 200$ mV, the peak position can be fitted by two straight lines: $-2.3\ RT/\alpha nF$ for the cathodic and $2.3\ RT/(1 - \alpha)nF$ for the anodic peak, respectively; Figure 2.12b. From these fits one can easily extract the value for the transfer coefficient α and then determine the value of the apparent electron-transfer rate constant k by the following equation:

$$\ln k = \alpha \ln(1-\alpha) + (1-\alpha)\ln\alpha - \ln(RT/nFv) - \alpha(1-\alpha)nF\Delta E_P/2.3RT \tag{2.41}$$

where R and F are the gas and Faraday constants, respectively.

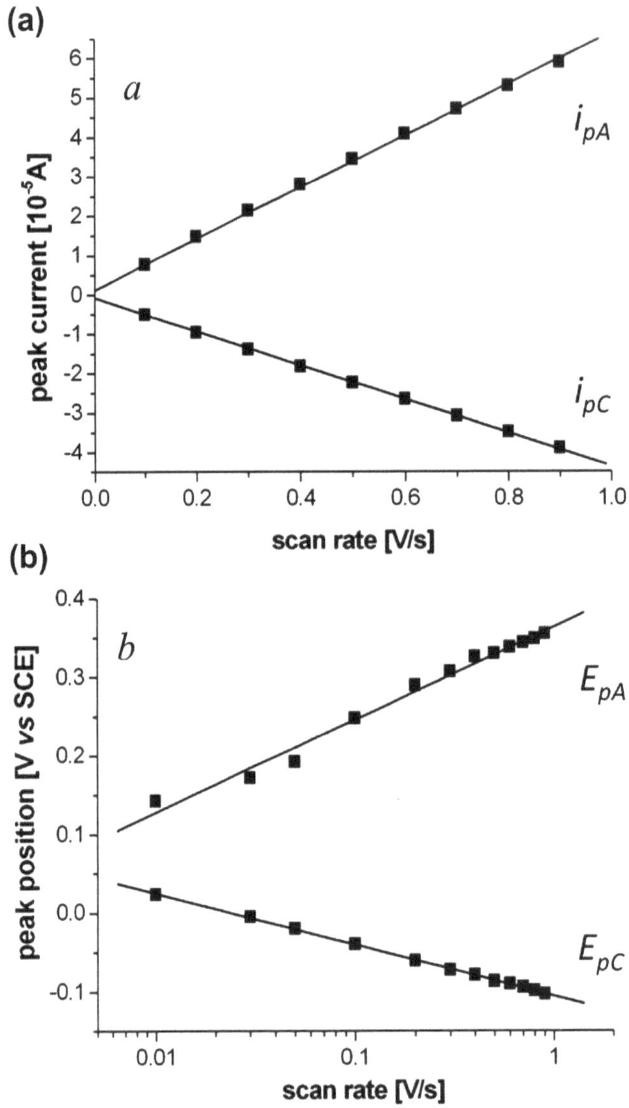

FIGURE 2.12

(a) Plot of i_{pA} and i_{pC} vs v and corresponding linear fit for hydroquinone derivatives immobilized on gold; (b) plot of E_{pA} and E_{pC} as a function of $\ln v$ and corresponding linear fit for $|\Delta E_p| > 200$ mV.

2.6.3 Electrochemical impedance spectroscopy

Studying electron reactions by LSV or CV implies imposing large perturbations on the system. In fact, the result of imposing a potential sweep on an electrode is to drive it far from equilibrium, while observing its (transient) response. Another possible approach, which will be the subject of the present section, is that of perturbing the electrochemical cell with an alternating signal of small amplitude and observing the way in which the system follows the perturbation at steady state. Such an approach has the advantages of (i) allowing for high-precision measurements, (ii) enabling theoretical treatments within the linear description of current-potential characteristics and (iii) enabling measurements over a wide range of frequencies (10^{-4}–10^{6} Hz) or time.

The experimental technique that allows one to retrieve cell or electrode impedance as a function of frequency is called electrochemical impedance spectroscopy (EIS).

As already mentioned, this method is capable of high precision and it is frequently used for the evaluation of charge transfer parameters and for the study of electrical double-layer structures, including the role of self-assembled monolayers at the WE surface. Usually it involves the use of an excitation AC signal of very low amplitude (order of magnitude of mV), so that data interpretation can take advantage of the linearity of current-overpotential relations in these conditions.

Theories have been developed to relate measured impedance parameters to chemically relevant quantities. In EIS these theories make use of equivalent circuits for describing cell electrical behavior.

2.6.4 Equivalent circuit of a cell

Generally speaking, an electrochemical cell can be regarded as an impedance to a small sinusoidal excitation. That is why it is conceivable to represent its performance as an equivalent circuit containing resistors and capacitors. One of these circuits, frequently used in interpreting EIS data, is the so-called *Randles equivalent circuit*, shown in Figure 2.13. It accounts for the fact that the total contribution to the current through WE is the sum of the faradaic process and the double-layer charging. The double layer contribution is almost that of a pure capacitance, thus it is described in the circuit by the capacitor C_{dl}. As to the faradaic contribution, its behavior is more complex and it has to be modeled as a frequency-dependent impedance, Z_f. The presence of the series resistor R_S accounts for the solution resistance through which any current in the cell has to pass. The faradaic impedance has been considered in the literature in various ways: among those, as a series of a solution resistance, R_{sol}, and a pseudocapacity, C_{sol}, and a pure charge-transfer resistance, R_{ct}, in series with a general Warburg impedance, Z_W, which accounts for a sort of resistance to mass transfer. Whereas usually R_S and C_{dl} behave as a real resistor and capacitor, respectively, the components of the faradaic impedance are frequency-dependent. It is a straightforward consequence that a given equivalent circuit is valid only for a certain frequency, but not for all. In the literature many different circuits have been proposed for different situations, and specific literature should be consulted to gain a deeper insight.

The approach known as EIS deals with the analysis of the variation, in the complex plane, of cell impedance as a function of ω. It therefore relies on Nyquist or Bode plots. In particular, the measured total impedance Z is expressed as $Z_{Re} + iZ_{im}$. Plotting Z_{Im} vs Z_{Re} for different values of ω, it is possible to extract chemical information from the EIS measurements. Solving the transfer function for the equivalent circuit of Figure 2.13 one obtains:

$$Z_{Re} = R_S + \frac{R_{ct} + \sigma\omega^{-1/2}}{\left(C_{dl}\sigma\omega^{1/2} + 1\right)^2 + \omega^2 C_{dl}^2\left(R_{ct} + \sigma\omega^{-1/2}\right)^2} \quad (2.42)$$

$$Z_{Im} = \frac{\omega C_{dl}\left(R_{ct} + \sigma\omega^{-1/2}\right)^2 + \sigma\omega^{-1/2}\left(\omega^{1/2}C_{dl}\sigma + 1\right)}{\left(C_{dl}\sigma\omega^{1/2} + 1\right)^2 + \omega^2 C_{dl}^2\left(R_{ct} + \sigma\omega^{-1/2}\right)^2} \quad (2.43)$$

where σ is a parameter whose value is given by the following relations linking the resistance and capacitance values of the faradaic impedance:

$$R_{sol} = R_{ct} + \sigma/\omega^{1/2} \quad (2.44)$$
$$C_{sol} = 1/\sigma\omega^{1/2}.$$

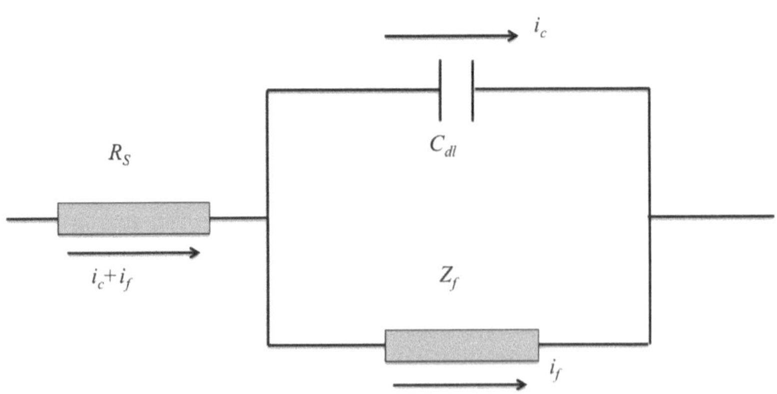

FIGURE 2.13

Randles equivalent circuit used in EIS data interpretation.

2.7 References

Bard, A. J., & Faulkner, L. R. (2001). *Electrochemical methods—fundamentals and applications* (2nd ed.). New York: John Wiley & Sons.

Grahame, D. C. (1953). Diffuse double layer theory for electrolytes of unsymmetrical valence types. *J. Chem. Phys., 21*, 1054–1060.

Hamann, C. H., Hamnett, A., & Vielstich, W. (1998). *Electrochemistry.* Weinheim: Wiley.

Hiemenz, P., & Rajagopalan, R. (1997). *Principles of colloid and surface chemistry.* New York: Marcel Dekker.

Jönsson, B., Wennerström, H., & Halle, B. (1980). Ion distributions in lamellar liquid crystals. A comparison between results from Monte Carlo simulations and solutions of the Poisson-Boltzmann equation. *J. Phys. Chem., 84*, 2179–2185.

Laviron, E. (1979). General expression of the linear potential sweep voltammogram in the case of diffusionless electrochemical systems. *J. Electroanal. Chem., 101*, 19–28.

Marcus, R. A. (1956). On the theory of oxidation-reduction reactions involving electron transfer. *I. J. Chem. Phys., 24*, 966–978.

Marcus, R. A. (1968). Electron transfer at electrodes and in solution: comparison of theory and experiment. *Electrochem. Acta, 13*, 993–1004.

Marcus, R. A., & Siddarth, P. (1992). Theory of electron transfer reactions and comparison with experiments. In E. Kochianski (Ed.), *Photoprocesses in transition metal complexes, biosystems and other molecules.* Amsterdam: Kluwer.

Verwey, E. J. W., & Overbeek, J. T. G. (1948). *Theory of the stability of lyophobic colloids.* Amsterdam: Elsevier.

CHAPTER

Life and the Water-Based Environment

3

CHAPTER OUTLINE

3.0 The peculiar chemical-physical properties of water

Liquid water is the most abundant liquid on Earth and for many aspects its relevance is self-evident, since it represents the main molecular component of the environment where life exists.

Biomolecular Electronics. http://dx.doi.org/10.1016/B978-1-4557-3142-8.00003-0

Before facing the extremely rich and complex role that water plays in biology, it is our intention to discuss the peculiar chemical-physical characteristics that water displays. These are, of course, intimately connected with its molecular structure, whose apparent simplicity contrasts with the difficulty in finding a satisfactory quantitative description, in terms of molecular models, of its amazing properties. We anticipate, moreover, that the particular interactions taking place in water, i.e., the hydrogen bond and the hydrophobic interaction, are at the basis of the role of this liquid as *the solvent* of life.

For a liquid of such a low molecular weight, water displays surprisingly high melting and boiling points and latent heat of vaporization. Of course, there exist many other substances of low molecular weight and high melting and boiling points, but these are invariably ionic crystals or metals whose structures are held together by strong ionic or metallic bonds. These properties suggest that there is present in water an intermolecular interaction that is stronger than that expected for ordinary, albeit highly polar, liquids.

The density maximum at 4 °C exhibited by liquid water, and the unusual phenomenon that the solid (ice) is less dense than the liquid (water), indicates that in the ice lattice the molecules prefer to be farther apart than in the liquid. Furthermore, one can conclude that the strong intermolecular bonds present in ice are also preserved in the liquid state and that they must be highly orientation-dependent, since water adopts a tetrahedral coordination (four nearest neighbors per molecule) rather than a higher packing density (such as, e.g., the 12 nearest neighbors characteristic of closely packed van der Waals solids where *bonds* are non-directional). Water shows other unusual properties (69 anomalous properties of water have been counted: http://www.lsbu.ac.uk/water/index2.html), such as a very low compressibility and peculiar solubility properties both as a solute and as a solvent.

However, peculiarities in chemical-physical properties are not limited to the liquid state, being even more marked in ice (Hobbs, 1974). It is, indeed, useful to start considering solid water to try to understand the origin of the liquid state's peculiar characteristics. The high molecular dipole moment and dielectric constant of liquid water may apparently be related via the Debye-Langevin and Clausius-Mossotti equations. But, at variance with any typical polar liquid whose dielectric constant decreases abruptly when it solidifies, due to the freezing of the molecular (i.e., dipolar) rotations, when water freezes, the dielectric constant actually increases, and it is still increasing at −70 °C (Hasted, 1973). It is very unlikely that such a phenomenon can be explained in terms of the conventional model of rotating dipolar molecules. The proton conductivity and mobility in ice are also higher than in the water (Kunst & Warman, 1980; Hobbs, 1974). Both these phenomena suggest that the ice lattice features some preferential pathways for the movement of charges, especially protons, via a "proton hopping" mechanism along the hydrogen-bond network (Pauling, 1935; Hollins, 1964). Such a mechanism appears to be present also in the liquid state (where also electron transfer appears to be assisted by water structure; Lin et al., 2005) and is known as the *Grotthus mechanism*. Thus, to understand the causes that endow liquid water with its unique characteristics it is useful to start by considering ice.

3.1 The hydrogen bond and the structure of water

The aforementioned unusual properties of water are connected to some sort of highly anisotropic interaction between molecules. In order to shed light on the nature of such interaction and on the type of bond involved, it is useful to consider the distances between various atomic centers in the ice lattice. Whereas the intra molecular O–H distance is about 0.10 nm, as expected for this covalent bond, the inter molecular O···H distance is only 0.176 nm, much less than the figure of 0.26 nm expected from summing the two van der Waals radii but still larger than the distance of 0.10 nm typical of a covalent bond. Such bonds, which apparently posses some covalent character, are known as hydrogen bonds, and the interested reader can find further in-depth information in the literature (Pauling 1960; Coulson 1961; Joesten & Schaad 1974; Jeffrey 1997; Schuster et al., 1976).

The hydrogen bond is special as it involves only a hydrogen atom, which, by virtue of its tendency to become positively polarized and its uniquely small size, can interact strongly with nearby electronegative atoms, resulting in an effective H-mediated "bond" between two electronegative atoms (such as O, N, F and Cl). Therefore, water is just a particular case among other hydrogen bond-forming substances.

Although the hydrogen bond is now believed to be a purely electrostatic interaction (Coulson, 1961), there is no simple equation for its interaction potential, at least not one that can provide quantitative predictions. One can find that the strengths of hydrogen bonds tend to follow a $1/r^2$ distance dependence, as in the case of the charge-dipole interaction.

Hydrogen bonds can occur both at intermolecular and intramolecular level and can exist in nonpolar environments. They are particularly important in macromolecular and biological assemblies, such as in proteins, linking different segments together inside the molecules, and in nucleic acids, where they are responsible for the structure of the DNA molecule.

Hydrogen bonds play a particularly important role in water; each oxygen atom with its two hydrogens can participate in four such linkages with other water molecules: two involving its own H atoms and two involving its unshared (lone-pair) electrons with other H atoms. To figure out how this is possible, one requires some picture of the charge distribution within the water molecule. Numerous models have been proposed (Ciccotti et al., 1987; Mahoney & Jorgensen, 2000; Guillot, 2002), but there is no single model that is able to account satisfactorily for the properties of water in all three phases (ice, liquid and vapor) and especially its most specific one, i.e., the hydrophobic interaction.

In most models the water molecule is described as having positive charges centered on each hydrogen atom and two negative charges on the opposite side of the oxygen atom, representing the two unshared electron pairs. Figure 3.1 shows as an example one of the many models introduced in the literature (Mahoney & Jorgensen 2000). The four charges are located along four arms radiating from the center of the O atom in a tetrahedral

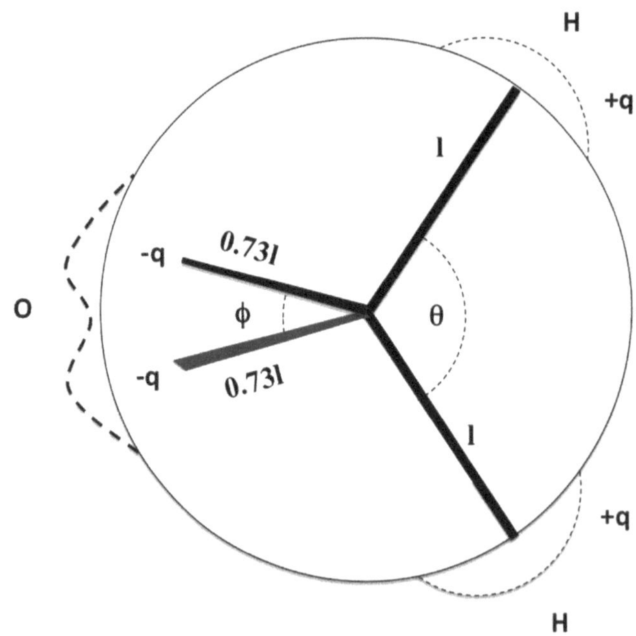

FIGURE 3.1

One of the most diffused models (TIP5P) for the water molecule (modified from Mahoney & Jorgensen, 2000). Here $q = 0.24$ e, $l = 0.096$ nm, $\theta = 104.5°$ and $\phi = 109.5°$.

configuration. The interaction between two water molecules is assumed to involve an isotropic Lennard-Jones potential and 16 Coulomb terms accounting for the interaction between each of the four point-charges on one molecule with the four on the other. The net Coulomb interaction depends on the mutual orientation of the two molecules. Computer simulations show that in ice each oxygen atom is tetrahedrally coordinated to four other oxygens, with a hydrogen atom lying along the line joining two oxygen atoms. This preferred linearity of the O–H···O bond in water is the main factor responsible for its strongly directional nature.

In liquid water the tendency to retain the ice-like tetrahedral H-bond network remains, but the structure is now disordered and labile. The average number of nearest neighbors per molecule increases to about five (hence the higher density of water on melting), but the mean number of H-bonds per molecule decreases to about 3.5, the lifetimes of which are estimated to be about 10^{-11} s.

It is also believed that the H-bond structure in liquids like water is cooperative in the sense that the presence of H-bonds between some molecules enhances their formation in nearby molecules, thereby tending to propagate the H-bonded network. The origin of this cooperativity seems to reside in the way that a well-structured network of H-bonds provides for the movement of protons, which increases the entropy of the system (Pauling 1935). Indeed, it is worth noting that the diffusion coefficient of protons in ice is significantly higher than in water.

Furthermore, it is also interesting to note that the tetrahedral coordination of the water molecule is also at the heart of the unusual properties of water, probably more than the hydrogen bonds themselves. Indeed, only a tetrahedral (or higher order) coordination allows for a three-dimensional network to form, providing a very large variety of different possible associations (Israelachvili 2005).

3.2 The hydrophobic effect

Strictly connected to water's tendency to make H-bond networks, there is the type of interactions that it displays when one considers it as a solvent or a solute, that is, when water interacts with molecules other than itself, especially nonpolar ones, that cannot form H-bonds (e.g., hydrocarbon chains). When water molecules come into contact with such molecules, it apparently seems that one or more of the four charges per molecule will have to face the non-polar solute, being unable to give rise to

H-bonds. Of course, the best configuration would be that having the least number of tetrahedral charges pointing toward the non-polar species, so that the other charges can point toward the water phase and are still available to be involved in H-bonding. Indeed, there are many ways for preserving H-bonds. If the nonpolar molecule is not too large, it is possible for water molecules to pack around it without giving up any of their hydrogen-bonding sites. Figure 3.2 shows a possible arrangement of water molecules around a quite small hydrophobic solute. Since the dispersion interaction between water and hydrocarbons is not very different from that between water molecules, the main effect of bringing water molecules and non-polar molecules together is the reorientation or restructuring of the water molecules so they can participate in H-bond formation more or less as in bulk water.

Thanks to the ability of tetrahedrally coordinated H-bonded molecules to link themselves together around almost any inert molecule, independently of its size or shape, the potential problem of having to give away some H-bonds

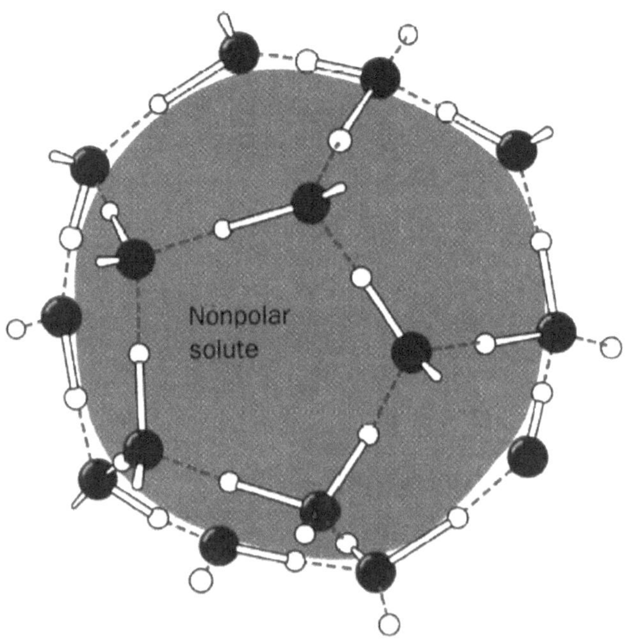

FIGURE 3.2

Possible re-organization of the water H-bond network around a small hydrophobic solute.

in the presence of inert molecules is often easily solved. Indeed, since water molecules in the liquid state participate on average in about 3.0–3.5 H-bonds, the result is that around certain inert solute molecules the water molecules can even have a higher coordination (of four), thus displaying an even lower enthalpy than in the bulk liquid.

Therefore, it is clear that the size and shape of nonpolar molecules are critical in determining the water structure adopted around them and that the structures adopted around small solute molecules may be very different from those adopted at larger ones. This phenomenon is often referred to as hydrophobic solvation or hydrophobic hydration. Unfortunately, there is no simple theory for such solute–solvent interactions. However, both theoretical and experimental studies indicate that the reorientation, or restructuring, of water around hydrophobic solutes or surfaces is entropically very unfavorable, since it disrupts the existing water structure and imposes a new and more ordered structure on the surrounding water molecules.

This point is a critical one since it is at the basis of the mechanisms driving biomolecules towards their native, functionally active configuration. In the following sections we will consider more explicitly the role that a solvent like water plays in structuring and driving the assembly of both single biomolecules and relevant biological structures.

Closely connected to the described hydrophobic effect is the resulting strong interaction taking place between nonpolar solute molecules immersed in water. This strong interaction in water cannot be accounted for by the conventional theory of van der Waals forces, which predicts, indeed, a reduced interaction in water. It is to be remarked that there is no bond associated with this mainly entropic phenomenon, which arises primarily from the rearrangement of H-bond configurations in the overlapping solvation zones as two hydrophobic molecules come together and is therefore of much longer range than any typical bond.

Given the complexity of the picture, there is to date no satisfactory theory of the hydrophobic interaction between solute molecules in water, even if a number of promising theoretical approaches have been proposed (Dashevsky & Sarkisov, 1974; Pratt & Chandler, 1977; Marcelja et al., 1977; Pangali et al., 1979; Luzar et al., 1987; Nicholson & Parsonage, 1982).

Nevertheless, experimental determinations of the distance-force law for hydrophobically interacting surfaces (Israelachvili & Pashley 1982) showed that the range of the hydrophobic interaction is much (6 to 10 times) longer than that typical of van der Waals interaction. This fact appears of extreme relevance in explaining the role of hydrophobic interaction in assembling biological structures.

3.3 The role of water in biology

A review by M. Chaplin, dated 2006, started with the following challenging sentence: "It is surely time for water to take up its rightful position as the most important and active of all biological molecules ..." (Chaplin 2006). This incipit provides the right setting for considering the role of water in biological phenomena.

Whereas it is generally clear that water plays a key role in life-as-we-know-it, to the extent that it is sought as the blueprint for the possibility of the presence of life even on extraterrestrial planets, its role, at the molecular level, has been often underestimated or considered ancillary to the prominent one played by the biological macromolecules themselves. The recognition of the role of water as one of the main players in biology refers exactly to its role not only as the *milieu* of choice for life to exist, rather as an essential ingredient for biological phenomena, reactions and systems to take place. In this sense, water should really be considered itself a biological molecule as in the case of any other biomolecule or biomolecular assembly that has its own native structure and functional properties thanks exactly to the involvement of water. Water not only drives and stabilizes molecular structures, but is actively involved in biomolecular functionality, providing, probably, the simplest, paradigmatic exemplification of *system biology*, where biological structure and function are not inherent to the single biomolecule, but rather to the macromolecule-water system. In other terms, it is meaningless to consider native biomolecular structure and function separately from the presence and the role of water since, as such, they would be simply non-existent.

Albeit a detailed analysis of the role of water in biomolecular systems is beyond the scope of this book, in the next sections we will focus on water's controlling influence over protein and nucleic acid structure and function, neglecting that over cellular activity. The interested readers will find exhaustive readings about the role of water in cellular activity in the literature (see, e.g., Chaplin 2004).

3.4 Water and biomolecules

As we have mentioned, water is an integral part of many biomolecules. In particular, water–protein interactions determine and shape the free-energy landscape that governs the folding, structure, stability and activity of proteins (Halle, 2004). In what follows we will see in some detail the role of water in these various contexts.

3.4.1 Protein folding

Proteins fold rapidly into well-defined three-dimensional shapes that constitute native structures. These depend on the primary structure, or sequence, of their amino acids. For this to be possible, the folding-energy landscape must consist of a multidimensional funnel with only small energy barriers that can be easily overcome by the available thermal energy. This requires a mechanism that involves flexible, exchangeable and extensible linkages, and water-mediated H-bonding is ideally suited for this purpose. Nonetheless, it has been proposed only recently that protein folding is mediated and guided by aqueous solvation (Cheung et al., 2002). Most protein folding takes place with water hydrating the peptide backbone and precisely maneuvering the backbone through its secondary structural assembly towards its ultimate, functionally active structure, the native one. Only in later phases of structural organization is water expelled by cooperative peptide H-bonding interactions from the more hydrophobic areas towards the more hydrophilic ones.

3.4.2 Protein structure

Water does not simply play the role of filling up the available space in and around proteins. Indeed, water molecules occupy specific sites and form localized clusters whose structure is determined by their hydrogen-bonding capabilities and by the structure of the protein. For example, they occupy cavities in most globular proteins, where they are present in similar amounts to the amino acids that are present and conserved as an integral part of the protein (Halle 2004). Although these water molecules are exchangeable with external ones, many of them are as essential for function as the amino acids. Surface water molecules are held in place most strongly by charged amino acids (Bandyopadhyay et al., 2005), but also by other polar groups. The exchange of surface water, and therefore the persistence of local clustering and the overall flexibility of the system, is controlled by the exposure of the water molecules to the bulk solvent. Greater exposure correlates with greater flexibility and freer protein-chain movement.

Water molecules can bind between protein areas that approach one another, albeit not touching, and can transfer surface information between subunits and towards the bulk. Being water molecules and their clusters extremely accommodating in terms of orientation and spatial distribution, they can lower energy barriers between the energy minima of neighboring systems. Water can therefore act as a "generalized" lubricant in that it facilitates the rearrangements of peptide amide–carbonyl hydrogen bonding during conformational changes, which gives proteins the flexibility they need for their functional activity (Fernández & Scheraga, 2003). The hydrogen bonds of water molecules and their clusters can also function as mechanical connections that transmit relative movement between different subunits and domains.

3.4.3 Protein activity and dynamics

A certain amount of water is necessary for the biological activity of all proteins. Most soluble enzymes, for instance, require a substantial number of water molecules to function. Recent work has indicated that, at the least, a spanning network of hydrogen-bonded water molecules that covers most of the surface of an enzyme is required for enzyme activity (Smolin et al., 2005). Such networks connect all of the hydrogen-bonded water clusters on the surface of a protein and control the dynamics of a protein, such as the motion of its domains. Reciprocal structural interactions between water molecules and a protein thereby transmit information around a protein (Nakasako, 2004). Such dynamics, in turn, underlie the biological function of a protein (Eisenmesser et al., 2005). It is worth noting that the spanning water network can influence protein dynamics. Experiments that have disrupted the spanning network, for example by changing the temperature to increase hydrogen-bond breakage, support this view. Such disturbances tend to cause a protein to denature and lose its biological activity, but only in the presence of water (Brovchenko et al., 2005).

The strength of the water network manages protein dynamics: greater protein movements take place in regions that contain many weak, bent or broken hydrogen bonds, due to the presence of more reactive water. On the other hand, restricted movements result from the presence of many strong aqueous hydrogen bonds. As a consequence, the structural elements that are most exposed to the bulk aqueous media are the least restricted (Bandyopadhyay et al., 2005). The water network links secondary-structure elements of proteins and therefore not only determines protein fine structural details, but also acts in selecting particular molecular vibrations (Smolin et al., 2005).

3.4.4 Protein–ligand interactions

Ordered water molecules at the surfaces of and between biomolecules frequently mediate binding interactions. Water-induced effects can be large and decisive in the selection of the best binding site (Ben-Naim, 2002). As in the case of folding, the displacement of bound water molecules from their surface hydration sites often compensates energetically for the formation of direct interactions between biomolecules. There is normally a close relationship between the enthalpic loss and entropic gain of the water molecules that are displaced and the enthalpic gain and entropic loss of the protein–ligand interactions that form, which facilitates both binding and release (Li & Lazaridis, 2005). However, water molecules can remain in place when a new interaction is formed, and they, together with peripheral hydrating linkages, can affect the specificity, stereochemistry and thermodynamics of the binding process (Ben-Naim, 2002). The energetic optimization of hydrogen-bonded networks (including their entropic consequences) that involve a protein, water and a ligand is an intrinsic part of all molecular recognition processes that involve binding proteins and enzymes.

3.4.5 Water and nucleic acid structure

The average strength of the interactions of water molecules with nucleic acids is much greater than that with proteins due to the highly ionic character of nucleic acids (Makarov et al., 2002). Although the importance of water is often overlooked, both DNA structure and the crucial recognition of its sequence rely on aqueous interactions, with the DNA helix expanding and contracting depending on its hydration status. The specific secondary structure assumed by the DNA double strand is hydration-dependent, possibly changing from the typical β (Watson-Crick type) conformation to an A-structure in the case of marked dehydration (such as that experienced by double-strand DNA in UHV-STM experiments; Driscoll et al., 1990). Water hydrates both the major and the minor grooves of DNA by forming links with the polar atoms at the edges of base pairs, and the orientation of the water molecules depends on the bases and their sequence. The hydration in the minor groove assumes a complex pattern that includes water hexagons in the initial spine of hydration, secondary hydration, and hydration up to the fourth aqueous shell (Arai et al., 2005). This sequence-dependent hydration, which extends to the rim of the grooves, can function as a "hydration fingerprint" for a given DNA sequence. This enables proteins to sense the base sequence from the outside of the grooves, and allows the rapid and lubricated sensing of the DNA sequence (Fuxreiter et al., 2005). The weak binding of proteins to DNA involves mainly the secondary hydration, which allows proteins to slide along the DNA in a manner that is facilitated by the remaining primary hydration water molecules. When specialized proteins encounter specific sites, they bind strongly to the DNA, and this is necessary for the processing of genetic information. Proteins do this by releasing and replacing some of the bound water molecules, which compensates for the entropic cost of protein binding.

3.4.6 Lipids and membranes

Another class of biological systems that depend primarily on the presence of a water-based environment for their assembly and physiological activity is that of membranes. These biological systems are the most common cellular structures in both animals and plants, where they are involved in many aspects of cellular activity, ranging from simple mechanical functions such as confining cellular material in a definite portion of space, cell motility, food entrapment and transport, to highly specific biochemical processes such as energy transduction, immunological recognition, nerve conduction and biosynthesis. Biological membranes are very complex and diverse. They commonly contain 50 or more different proteins and a number of phospholipids and glycolipids with various headgroups, numbers of chains, chain lengths and degrees of unsaturation, as well as steroids (e.g., cholesterol) and pure hydrocarbon molecules (e.g., squalane). In spite of their complexity, there are many aspects of membrane structure that may be qualitatively described in terms of the concepts we have already outlined; so it is useful to start with a consideration of membrane lipids.

These are basically amphiphilic molecules and, as such, their mutual interaction in a polar solvent like water is mainly determined by the combined action of the hydrophobic effect (see section 3.2) and related interaction on the one hand, and by the role of hydration of the polar or charged portions of the molecules, on the other hand.

Whereas the hydrophobic effect gives rise to a long-range, attractive interaction between the apolar molecular moieties, the role of hydration is opposite in sign. Indeed, its effect can be recognized in the tendency of certain molecules and groups to be water-soluble and to strongly repel each other in water.

It is from the interplay between the attractive forces developing between the hydrophobic parts of lipid molecules (i.e., their chains) and the repulsive interactions occurring between the hydrated, hydrophilic headgroups that structures like biological membranes and lipid bilayers tend to form and to be maintained in a water-based environment.

There are, however, overall structural requirements on the molecules to give rise to planar lipid bilayers. Self-assembling amphiphiles can be classified according to their "packaging properties" that determine the kind of molecular aggregate they can give rise to. These properties depend on lipids' optimal area a_0, the volume v of their hydrocarbon chain or chains, which are assumed to be deformable but incompressible, and the maximum effective length that the chains can assume, l_c.

This length sets a limit on how far the chains can extend; smaller extensions are allowed, whereas further extensions are not, these being energetically or entropically unfavorable. It is, indeed, a semiempirical parameter, since it represents a sort of cutoff distance beyond which hydrocarbon chains can no longer be considered as fluid.

Once the optimal surface area a_0, hydrocarbon chain volume v, and critical length l_c are specified for a given molecule – all these being measurable or possible to estimate – one may ascertain which structures the molecules can pack into within these geometric constraints.

A convenient parameter for analyzing these structures is the dimensionless number, $v/a_0 l_c$, known as the critical "packing parameter" or "packing factor". A molecule can assemble in a bilayer if its packing parameter is $\cong 1$. Lipids that fulfill these requirements, such as phospholipids, have two chains and a rather small headgroup that make them assume the shape of a truncated cone. The presence of two acyl chains contributes to enhancing the overall molecular hydrophobicity by lowering drastically their CMC (critical micellar concentration) (10^{-6}–10^{-10} M, in contrast to 10^{-2}–10^{-5} M for micelle-forming lipids), thus stabilizing the self-assembled structure to the extent that bilayers do not dissolve even in the absence of free lipids in the solution bulk. Furthermore, the lifetime of a molecule in an aggregate is also enhanced, along with a slow-down of the typical flip-flop time (10^2–10^5 s), which affects a number of the functional aspects of biomembranes.

3.5 Biological reactions taking place in water

Water is the medium where most biological processes take place. As such, it can change and affect them in an instant, slowing them down at one moment, speeding them up at another, while constantly and rapidly transferring information from one place to another, both near and far. The role of water can be exerted at different levels: at that of individual isolated molecules, of small clusters, of much larger networks or of liquid phases that can have different characteristics.

We have seen in the previous sections the role of water in affecting many properties of biomolecules and biosystems. Now, we shall consider its mediating role in biological reactions. Albeit any biological reaction takes place in water, we are, hereby, particularly interested in biological reactions and phenomena involving the action of electric fields and the consequent flow of charges under the form of electrons or ions. Therefore, we will proceed with a consideration of the electrical conduction in water, as far as it is relevant to biomolecular reactions.

Ordered water molecules that connect donor and acceptor sites in a charge-transfer pair can facilitate both proton-transfer processes and electron-transfer reactions. In both cases, the transfer is faster when the linking water molecules are more strongly hydrogen-bonded.

3.5.1 Water and proton transfer

There exists an extraordinary process, usually referred to as the Grotthuss mechanism, whereby protons can move speedily in an aqueous environment or in associated liquids in general (Cukierman, 2006).

Because transfer occurs between pools that contain low concentrations ($\sim 10^{-7}$ M) of protons, conventional diffusion (i.e., depending on concentration gradients) would be extremely slow. By contrast, the Grotthuss mechanism is most efficient in neutral solutions (Cukierman, 2006). Following the addition of a proton to one end of a hydrogen-bonded chain (often referred to as a "water wire", in analogy with the terminology used for electrical conductors), hydrogen-bond flipping down the chain releases a different but identical proton at the other end of the water wire. Protons therefore move fastest when an ordered one-dimensional hydrogen-bonded chain facilitates their movement (Cukierman, 2006). A common structural theme among proteins that are involved in proton translocation is a string of water molecules in a hydrophilic cavity that is lined with polar amino acids. Thermal fluctuations involving these polar amino acids modulate proton transfer by rocking the water molecules. So, the polarization of the hydrogen bonding varies with the closing and opening of the hydrogen-bonded distances, which facilitates proton transfer. In aquaporin molecules, which allow the single-file translocation of water molecules through membranes, halfway through their water-transfer channels water molecules are reoriented. This appears to be useful in preventing sequential hydrogen bonding and the resulting facilitated proton transfer through the membrane (Tajkhorshid et al., 2002).

The ability of water to assemble water wires and modulate their structure and extent is also the basis of its role in effecting electron transfer.

3.5.2 Water and electron transfer

Water can influence the electron-transfer (ET) reaction rates by mediating ET pathways, as well as by controlling free energies of activation

(Gray & Winkler, 2003; Marcus & Sutin,1985). In the recent past, the distance dependence of water-mediated ET reaction rates has been intensively investigated both experimentally and theoretically. Until recently, experimental and theoretical analysis suggested a single-exponential decay of the ET rates with distance through water, with a characteristic decay constant of about 1.6 to 1.7 $Å^{-1}$ (Gray & Winkler, 2003; Wenger et al., 2005; Ponce et al., 2000). In comparison with proteins,which exhibit inverse decay lengths of about 1.0 to 1.2 $Å^{-1}$ (Gray & Winkler, 2003), water appeared to be a rather poor ET mediator because of extensive through-space connections in tunneling pathways (Wenger et al., 2005).

Indeed, protein ET reactions play a key role in biologically relevant processes in living cells, most notably photosynthesis and respiration (Berg et al., 2002). It is particularly challenging to describe the structure dependence of intermolecular ET reactions because of the wide range of accessible docking geometries, and several studies have addressed these reaction mechanisms. The factors that control unimolecular ET rates, namely the donor-to acceptor (D to A) distance and energies, the structure of the ET-mediating protein medium and the thermal atomic motion, have been extensively explored both experimentally and theoretically.

Particularly, in the past few years, a number of important experimental observations have showed inconsistency with the single exponential decay model for water and with the generality of rapid distance decay for ET through water molecules. In crystals, ET across thin aqueous interfaces was found to be easily done (Tezcan et al., 2001). In covalently cross-linked azurin complexes (section 4.1.1), water dimers that formed between the redox centers appeared to increase the ET rate substantially (van Amsterdam et al., 2002). In DNA, the influence of water (and counterions) may be even more pronounced, and fluctuations in hydration appear to gate ET. In small water clusters in gas phase or at the surface of TiO_2, hydrated electrons appeared to facilitate water-mediated ET. These experiments all point to the need for a deeper unifying theoretical framework to describe the distance and structure dependence of water-mediated ET reaction rates.

Such a treatment was developed by D. Beratan and collaborators (Lin et al., 2005), who used computation to explore the way the aqueous protein environment influences ET rates as a function of distance. Their results show that the intervening water structure leads to one of three distinctly different ET tunneling regimes, at variance with the common assumption of single exponential distance decay (Gray & Winkler, 2003, 2005; Tezcan et al., 2001). The identification of these three regimes provides a framework for understanding the mechanisms that underlie several unexplained and seemingly unrelated water-mediated biological ET rate processes (Tezcan et al., 2001; van Amsterdam et al., 2002), as well as yielding a strategy for making theoretical predictions of bimolecular rates that take the mediating role of water into account.

Particularly, they calculated the mean-square electronic coupling $\langle T_{DA}^2 \rangle$ (section 4.5) dependence on distance between two porphyrin rings of trypsin-solubilized bovine liver cytochrome b_5 (Protein Data Bank (PDB) entry 1CYO) in a self-exchange ET reaction for the two systems proteins and water and protein-only mediation. From the analysis of their results, in agreement with experimental evidence, it turned out that only in the case where mediating water was present (i) did there exist multiple tunneling mediating regimes, and (ii) that the structured water coupling intermediate regime may provide an important mechanism to facilitate ET reactions in the critical near-contact distance range relevant to biological ET kinetics.

Thus, water may act as a particularly strong tunneling mediator when it occupies a sterically constrained space between redox cofactors with strong organizing forces that favor a constructive interference of coupling pathways.

It is perhaps worth recalling that a previous work by S. Lindsay and collaborators (Vaught et al., 1995) had studied the behavior of the conductivity of water in the tunneling gap between a substrate (Au(111)) and a tip (Pt/Ir) of an ECSTM (see section 4.5.1), finding a highly non-exponential trend.

3.6 Biological reactions and phenomena involving the action of electric fields

Charges are present everywhere and the biological environment is not an exception. Therefore, all biological phenomena and reactions take place under or are directly affected by the action of electric fields. Consider, for instance, the role of charged amino-acids around the catalytic cage of an enzyme that drive and stabilize the ligand in the catalytic site; or the role that charges play in forming stable interaction pairs between an antibody and its corresponding antigen. At the cellular level, the presence of charges on the surface of red blood cells, for example, prevents them from clumping together in normal conditions; in contrast, when charges on surface receptors are removed by the action of proteolytic enzymes, that repelling action ceases, enabling the formation of clots as, for instance, in the presence of a wound. Surface charges and their modulation are also responsible for phenomena such as membrane fusion. A typical example is the fusion of insulin-containing vesicles with the cytoplasmic membrane of β-cells in the pancreas, causing the release of insulin. Lastly, we recall the role of charge imbalance resulting from ion dissociation and selective membrane permeation, in building up the transmembrane potential difference present in any cell. From these few examples it is clear that the list of biological phenomena and reactions that involve an active role of charges and related electric fields in biology is nearly endless. As an instructive and useful example, we will consider in detail the case of voltage-gated ion channels, with a particular emphasis on the role of charges and electric fields in operating the gating of the channel.

3.6.1 Voltage-gated ion channels

Phospholid bilayer-based biological membranes are insoluble in and impermeable to water and ionic and charged moieties because of their highly hydrophobic core, made essentially of acyl chains. Nevertheless, life is to a large extent based on the possibility of the effective control of ion fluxes across the biological membrane. Generally speaking, the bi-directional exchange of water, molecular effectors, cations and anions through the membrane is at the basis of a number of key phenomena related to cellular homeostasis and information transfer among different cells.

Living beings have developed various approaches to control membrane permeation by non-lipophilic substances, among which ion channels are some of the most important and intriguing solutions. They are complex transmembrane proteins that assist selective permeation of the membrane by ions. Their functional activity is at the basis of the existence of the transmembrane voltage, i.e., the potential difference between the inner and outer cellular space, which in mammalian cells is typically around −60 mV (the electrophysiological convention requires that the membrane potential be always expressed as its value *inside* the cell with respect to that *outside*). Its rise is due to one of the most important characteristics of ion channels, that is, their selectivity, i.e., their ability to allow only specific ions (e.g., K^+ or Na^+ or Ca^{2+} or Cl^-) to pass through their permeation pore, when open. For sake of completeness, we must say that ion channel selectivity is not perfect; in the case of potassium ion channels, for instance, selectivity is typically 30:1, i.e., on average, for every 30 K^+ ions a different ion, typically Na^+, can flow through the channel. This fact is accounted for in the equation for transmembrane potential (vide infra) by a leakage term. Ion channels are passive transporters, that is, ion permeability through them is only driven by the concentration gradient (at variance with what happens in ion pumps, the other important family of ion transporters, which use energy for pumping ions against their concentration gradient).

The intracellular space has a higher concentration of K^+ than the extracellular space. Therefore, if for any reason a channel that is specific for K^+ opens, potassium ions will tend to flow outside the cell enevelope following their concentration gradient. Due to channel specificity, counterions (anions in this case) will be unable to follow potassium ions. The result is a charge imbalance between the inner and the outer cell membrane surface. The cell membrane acts, thus, as a parallel-plate capacitor whose inner armature is negatively charged, giving rise to a membrane potential according to the Nernst equation.

This equation can be derived by considering the net ion flow, j, through a channel as a function of chemical and electrical gradients across the membrane. If x defines the direction perpendicular to the plane of a membrane, one can write

$$j = -D\left(\frac{dC}{dx} + \frac{zF}{RT}C\frac{dV}{dx}\right) \tag{3.1}$$

where D is the ion diffusion coefficient, C the ion concentration, z the ion valence number, F the Faraday constant, R the gas constant, T the absolute temperature and V the transmembrane potential. When the net flow equals zero, i.e., $j = 0$, one can integrate equation (3.1), obtaining

$$V = \frac{RT}{zF}\ln\frac{C_o}{C_i} \tag{3.2}$$

which relates the potential V of an ion in equilibrium with the concentration gradient established by internal C_i and external C_o ion concentrations. This equation is known as the Nernst equation and it is usually rewritten by defining the equilibrium potential of ion N as E_N and its internal and external concentrations as N_i and N_o:

$$E_N = \frac{RT}{zF}\ln\frac{N_o}{N_i} \tag{3.3}$$

There are several different kinds of ion channels that are characterized by their specificity for a given ion and by the way they function. In particular, they can differ due to their gating mechanism, which can be due to a pH gradient, to the interaction with a given ligand, to mechanical stress and, most notably, to a change in membrane potential. The ion channel superfamily whose gating is caused by changes in membrane voltage is that of voltage-gated ion channels. These channels are very important because they are typically, even if not exclusively, found in the axons of neural cells and their functioning is at the basis of the generation and propagation of the nervous signal: the *action potential*.

The first voltage-dependent ion channel that was isolated and purified was extracted from the eel electroplax, where a large concentration of Na^+ channels is present (Agnew et al., 1978). Several years later, the sequence of the eel Na^+ channel was inferred from its mRNA (Noda et al., 1984). The first K^+ channel sequence elucidated was deduced from the Shaker mutant of *Drosophila melanogaster* (Tempel et al., 1987). These sequences formed the basis of further cloning of a large number of Na^+, K^+ and Ca^{2+} channels from many different species. A common pattern emerged from all these sequences: the functional channels are made up of four subunits (in the case of K^+ channels) or one protein with four homologous domains (for Na^+ and Ca^{2+} channels). Each domain or subunit has six transmembrane segments (α-helices) and a pore loop. The fifth and sixth helices (S5 and S6) and the pore loop are responsible for ion conduction. The fourth transmembrane segment (S4) contains several positively charged residues, arginines or lysines, and was initially postulated to be the voltage sensor (Noda et al., 1984). In addition S2 and S3 contain acidic residues such as aspartate and glutamate. Most of the channels have additional subunits that modify the basic function but they are not essential for voltage sensing and ion conduction.

One can think of voltage-dependent channels as comprising three basic parts: the voltage sensor, the pore or conducting pathway and the gate. The pore and the gate are located in the S5-loop-S6 region and the voltage sensor in the S1−S4 region. As the conduction depends upon the voltage across the membrane, one can draw an analogy with a field effect transistor. If we consider a typical *V*-dependent K^+ channel, its voltage sensor corresponds to the gate of a *p*-channel FET transistor, the conducting pathway of the ion channel corresponds to the *p*-channel and the gate of the channel is the space charge in the *p*-channel. As we will see below, this analogy is useful to discuss the parts of the channel but it cannot be pushed very far because, although the functions are similar, the actual mechanisms are quite different.

Living cells in general, and neurons in particular, are surrounded by a thin membrane made of a lipid bilayer. The permeability of ions through the lipid bilayer is extremely low because a large amount of energy is required to put a charged ion species inside a lipid bilayer due to its low dielectric constant (Parsegian 1969). The conducting pathway of ion channels accomplishes the function of lowering that energy barrier by providing a favorable local environment, thus allowing large ion flows under an appropriate driving force. The driving force responsible for ion flow corresponds to the difference between the voltage applied, V, and the voltage at which there is no flow, or reversal potential. If the channel is perfectly selective to just one type of ion, say K^+, then E is the Nernst potential; otherwise, E is predicted by the Goldman-Hodgkin-Katz equation, which considers concentrations and relative permeabilities (Hille, 2001). Knowing the conductance of the conducting pathway γ, one can compute the current flow i through the open conducting pore as:

$$i = \gamma(V - E). \tag{3.4}$$

The i-V curve for an open channel may be nonlinear because, in general, γ depends on voltage.

The ion conduction through the pore may be inhibited by closing a gate. Generally speaking, thermal fluctuations can close and open the gate randomly and the current displays interruptions. In voltage-dependent channels, the probability that the gate is open, P_o, depends on the membrane potential. In the majority of V-dependent Na^+, K^+ and Ca^{2+} channels from nerves and muscles, P_o increases with membrane depolarization (i.e., decrease in the resting potential). There are a few cases, such as that of the Kat1 channel (Hoshi, 1995), where P_o increases on hyperpolarization. The operation of the gate can be studied by recording the current flowing through a single ion channel (at a time). This is possible with the patch clamp technique (Hamill et al., 1981), which records currents from a very small patch of membrane with a small glass pipette and a low-noise amplifier that can resolve currents of the order of 1 pA.

As the internal concentration of K^+ is more than 10 times that in the extracellular space, the reversal potential E for K^+ channels is around −80 mV. Starting with a negative membrane potential of −100 mV, the channel is closed most of the time. A depolarizing voltage pulse to −30 mV increases the open probability and the channel spends some time in the open state.

As we are dealing with one molecule, thermal fluctuations will generate different responses for each repetition of the same pulse (stochastic behavior). A larger depolarization of +30 mV increases the P_o even more by increasing the open times and decreasing the closed times. The time elapsed between the start of the pulse and the first opening (first latency) is decreased for the larger depolarization. Apart from increasing the open times, the magnitude of the current through the pore is increased by a larger depolarization. This is because the V applied is now much higher than E, increasing the driving force for ion movement. Thus, this increase in current is *not* a result of increasing P_o but is just a passive property of the open pore. An average of several thousands of repetitions gives us the macroscopic ionic currents. Provided the channels do not interact, the average of thousands of repetitions is the same as having thousands of channels operating simultaneously (hergodic system). The turn-on kinetics is faster for a more positive potential and the current magnitude is also increased. The kinetics change is the result of an increased P_o while the magnitude increase is the result of both increased P_o and driving force. One can now write an expression for the macroscopic current as

$$I = P_o(V,t)N\gamma(V - E) \tag{3.5}$$

where N is the channel density and $P_o(V, t)$ is the voltage- and time-dependent open probability.

How does P_o become voltage-dependent? It is clear that to detect changes in membrane potential, a *voltage sensor* is required. The electric field in the bilayer could be detected by electric charges or dipoles that change their position or orientation according to changes in the field. We will see below that the actual charges involved in voltage sensing have been identified. In the resting (hyperpolarized) condition, the membrane is negative inside and the positive charges are located in contact with the interior of the cell. Upon depolarization, the positive charges are driven outwards. This movement has two consequences: it is coupled to the gate, resulting in pore opening, and the charge translocation produces another membrane current, transient in nature, called gating current. It is called gating current because it ultimately gates the channel open and closed, and it is transient because the charge locations are bound to limiting positions as they are tethered to the protein.

To understand the voltage sensor, a characterization of the gating charge movement and a correlation of that movement to structural changes in the protein are thus required.

At this point, it is interesting to address two functional questions. The first question is what the kinetics and steady-state properties of the gating charge movement are and how this charge movement relates to channel activation. The second is how many elementary charges move in one channel to fully activate the conductance and how this movement takes place in one channel.

The movement of charge or dipole reorientation is at the basis of the voltage-sensor operation and was predicted by Hodgkin and Huxley (Hodgkin & Huxley, 1952). Gating currents are transient and take place only in the potential range where the sensor responds to the electric field; therefore, their behavior resembles that of a nonlinear capacitance. In addition, as gating currents are small, to record them it is necessary to decrease or suppress the ionic currents through the pore and eliminate the normal capacitive current required to charge or discharge the membrane. This is, normally, accomplished in a differential fashion by applying a pulse in the voltage range that activates the current and then subtracting the current elicited by another pulse or pulses in the voltage range that does not activate the currents to eliminate the linear components. Using these subtraction techniques, the kinetics of Na^+ gating currents were studied in detail in squid giant axon and other preparations where a high channel density was found. The combination of gating currents, macroscopic ionic currents and single-channel recordings was used to propose detailed kinetic models of channel operation (see, for instance, Vandenberg & Bezanilla, 1991). In the case of voltage-dependent channels cloned and expressed in oocytes or cell lines, it was possible to achieve large channel densities and study those channels in the virtual absence of currents from other channels. In comparison with the currents in natural tissues such as the squid giant axon, the expression systems' gating currents were much larger and made the recording easier and cleaner. In addition, the study of the pore region gave researchers the possibility of mutating the channel protein to eliminate ionic conduction but maintaining the operation of the gating currents. It is convenient to illustrate the basic features of gating currents and their relation to ionic currents in recordings from Shaker K^+ channels with fast inactivation removed (Shaker-IR).

Several features that are characteristic of most voltage-dependent channels can be observed. First, the ionic currents do not show significant activation for potentials more negative than −40 mV, while the gating currents are visible for all the pulses applied, suggesting that there is charge displacement in a potential region where most of the channels are still closed. Second, the time course of activation of the ionic current is similar to the time course of decay of the gating current. Third, the time course of the return of the charge (gating current "tail") changes its kinetics drastically when returning from a pulse more positive than −40 mV, which is precisely the potential at which ionic currents become clearly visible. The gating tails are superimposable for potentials more positive than −20 mV, showing that most of the charge has moved at −20 mV. The total charge moved at each potential may be computed as the time integral of the gating current for each pulse. As we will see below, it is possible to estimate the total charge moved per channel molecule; therefore, the voltage dependence of the charge moved can be plotted. Knowing the number of channels present (see below), using equation (3.5) it is possible to estimate the voltage dependence of P_o. A striking feature of the relation giving $Q(V)$ is that it is displaced to the left of the curve $P_O(V)$ so that there is quite a large charge movement in a region where the P_o is essentially zero. This is an expected feature of a channel that requires several processes to occur to go from closed to open, such as the classical Hodgkin and Huxley model, where four independent particles are needed to be simultaneously in the active position for the channel to be open.

Current recordings of the type described above can be used to formulate kinetic models of channel gating. These models are normally written as a collection of closed and open states interconnected by rate constants that, in general, depend on membrane potential. Initially, kinetic models were developed from the macroscopic ionic currents only. The addition of single-channel recordings and gating current recordings imposes several constraints on the possible models and on the parameters fitted, thus producing a more robust description of the kinetic parameters that characterize the channel.

In kinetic models it is a common assumption that the movement of the charge or dipole in the field has a finite number of low-energy positions separated by energy barriers.

According to transition state theory, the transition rates across the energy barrier are exponentially related to the negative difference of the free energy amplitude of the barrier. This free energy contains nonelectrical terms, and an electrical term such that the membrane voltage can either increase or decrease the total energy barrier, resulting in changes in the forward and backward barrier crossing rates. In the literature, there are several examples describing the kinetic and steady-state properties of various types of voltage-dependent channel (see, for example, Vandenberg & Bezanilla 1991).

A more general approach to modeling is based on a representation of a landscape of energy using the charge moved along a chosen reaction coordinate.

When gating currents are recorded with high bandwidth, new components are observed. Recording gating currents with a bandwidth of 10 kHz, before an initial plateau and decay of the gating current, there is a brief surge of current. When the bandwidth is increased to 200 kHz, the first surge of current is the predominant amplitude. The large spike of current is followed by an extended plateau that corresponds to the dip present at 10 kHz. However, the area spanned by the large spike is a very small fraction of the total charge moved during the entire time course of the gating current. The interpretation of this early component is best achieved if one uses the representation of the gating charge moving in a landscape of energy that undergoes a change in tilt when the membrane potential is changed such that the charge advances in its initial energy well before making the jump across the first energy barrier.

Indeed, one can learn a lot from these kinetic models. It is now clear that voltage-dependent channels have multiple closed states and, in some cases, can have several open states. In general, the opening of the gate requires all four subunits to be activated. Each subunit undergoes several transitions before reaching the active state, and, in the case of K^+ channels, they do not seem to interact until the final step that opens the channel. However, in the muscle Na^+ channel, site-directed fluorescence studies have shown that the interdomain interactions are manifested prior to channel opening. Kinetic modeling has provided us with a picture of the channel in terms of channel physical states with transitions between them that are dependent on voltage. Kinetic modeling is a critical step in developing a physical model of channel operation because all the predicted features of channel function should be reproduced by the structure of the protein and its voltage-induced conformational changes.

When a gating charge moves within an electric field, one detects a current in an external circuit. The time integral of that current gives the charge moved multiplied by the fraction of the field it traverses; therefore, our measurement of gating charges does not represent the exact number of charges displaced because it includes the arrangement of the electric field. One must keep this in mind when representing the reaction coordinate of the activation of the channel in the variable q. A channel evolves from $q = 0$ to $q = z_T$ traversing many closed and/or open states.

Activation of the channel corresponds to the opening of the pore and, in formal analogy to the case of chemical potential, it is possible to define the activation potential as

$$W_a = -k_B T \ln P_o \tag{3.6}$$

Then the activation charge displacement corresponds to the negative gradient of the activation potential

$$q_a = -\frac{dW_a}{dV} = k_B T \frac{d \ln P_o}{dV} \tag{3.7}$$

The equilibrium probabilities in each physical state of the channel can be explicitly written using the Boltzmann distribution knowing the potential of mean force F_i for each state i. Then, by associating open or closed (or intermediate states) conductance to every state, one can write an expression for P_o that includes the voltage dependence of F_i. The final result of the derivation gives a relation between $q(V)$, q_a and z_T and the charge moving between open states q_l:

$$q(V) = z_T - q_a - q_l. \tag{3.8}$$

This result is general and includes cases with any number of open and closed states connected in any arbitrary fashion. If there is no charge movement between open states ($q_l = 0$), then the $Q-V$ curve superimposes on q_a. In addition, it is possible to estimate z_T, the total charge per channel, by taking the limiting value of q_a that makes $q(V)$ equal to zero. In the typical case of a channel that closes at negative potentials, one obtains:

$$Z_T = \lim_{V \to -\infty} k_B T \frac{d \ln P_o}{dV} \tag{3.9}$$

a result that was first obtained for the special case of a sequential series of closed states ending in an open state. This method has been applied to several types of voltage-dependent channels, and the charge per channel was found to range between 9 and 14 e_0 (Hirschberg et al., 1996; Noceti et al., 1996; Seoh et al., 1996).

Another possible way to assess the total charge per channel is to measure the maximum charge from the $Q-V$ curve and divide by the number of channels present. The number of channels can be estimated by noise analysis or by toxin binding. The value of charge per channel estimated by the method was 12–13 for the Shaker K^+ channel, a figure similar to that obtained by the limiting slope method (Seoh et al., 1996). As the limiting slope measures only the charge involved in opening the channel, the agreement between the two methods implies that in the case of the Shaker channel there is no peripheral charge.

The large value of 12 to 13 e_0 per channel explains the very steep voltage dependence of the superfamily of voltage-gated ion channels. At very negative V, q_a has a linear dependence on V, so that P_o is exponential in V, such that it increases by e in only 2 mV,

$$P_o \infty \exp(z_T V / k_B T). \tag{3.10}$$

There are many ways one could envision how the charge movement is produced by the channel protein. The putative α-helical transmembrane segments have an intrinsic dipole moment that, upon tilting in the field, would produce an equivalent charge movement. Also, induced dipoles of amino acid side chains could account for the same effect. However, 13 e_0 per channel is very large, and charged amino acids become the most likely scenario. Since the first channel was cloned, it was recognized that the S4 segment with its basic residues would be the prime candidate as the voltage sensor. By introducing mutations that neutralize the charges in S4, several studies found that there were clear changes in the voltage dependence of the conductance. It is interesting to note that in several instances a neutralization of one particular residue decreased the total gating charge by more than 4 e_0. This indicates that somehow the charges interact with the electric field where they are located such that the elimination of one charge can affect the field experienced by the remaining charges. If most of the gating charge is carried by the S4 segment (4 e_0 per subunit), it gives a total of 16 e_0. Therefore, to account for the 13 e_0 for the total channel obtained from charge/channel measurements, they all must move at least 81% of the membrane electric field ($16 \times 0.81 = 13$).

A 3D structure of the channel, even in only one conformation, would be invaluable as a guideline in locating the charges inside the protein and in proposing the other conformations that account for the charge movement consistent with all the biophysical measurements. The long-awaited first crystal structure of a voltage-dependent channel, KvAP from archea *Aeropirum pernix*, was published in 2003 (Jiang et al., 2003a). The structure was a surprise because it showed the transmembrane segments in unexpected positions with respect to the inferred bilayer. For example, the N-terminal was buried in the bilayer whereas it has been known to be intracellular; the S1−S2 linker is also buried although it has been previously shown to be extracellular. The S4 segment, along with the second part of S3 (S3B), formed the *paddle structure*, which was intracellular and lying parallel to the bilayer, a location that would be interpreted as the closed position of the voltage sensor.

However, the pore gates in the same crystal structure clearly correspond to an open state. Thus, the crystal structure is in a conformation that was never observed functionally, raising the question whether that crystal structure of KvAP is indeed representative of the native conformation of the channel in the bilayer. The authors mention that the channel is inherently floppy; therefore, in obtaining the structure, it was necessary to co-crystallize it with F_{ab} fragments (see Chapter 5) attached to the S3−S4 loop of the channel. This raises the possibility that the fragments may have distorted the channel structure that was in the final crystal. The authors functionally tested the structure by incorporating the channel in bilayers and recording the currents through the channel after F_{ab} fragments had been added to the inside or to the outside of the channel. The F_{ab} fragment did not attach from the inside but only from the outside, showing that the position of the S3−S4 shown in the crystal structure (obtained in detergent) did not represent a native conformation of the channel in the bilayer (Jiang et al., 2003b). Along with the crystal structure of the full KvAP channel, Jiang et al. (2003a) solved the crystal structure of just the S1 through S4 region of KvAP. This crystal showed similarities but was not identical to the S1−S4 region of the full KvAP crystal structure. This second crystal was docked to the pore region of the full crystal maintaining the S2 segment within and parallel to the bilayer to obtain a new structure that was proposed in the open and closed conformations as the paddle model of channel activation.

As to the model of sensor movement, three different classes of models have been proposed to explain the charge movement in voltage-dependent channels. The charge that moves resides completely in the first four charges of the S4 segment. The common feature of all three models is that the charge is translocated from inside to outside upon depolarization. However, there are important differences as to how those charges relocate in the protein structure.

The first one is called the helical screw model. Although there exist different versions of this model, the general idea is that upon depolarization the S4 segment rotates along its axis and at the same time translates as a unit perpendicular to the membrane, thus changing the exposure of the charges from the intracellular to the extracellular solution, effectively translocating 4 *e* per subunit. In the original version of the model, the change of exposure required a large 16-*e* translocation of the S4 segment, and the positively charged arginines were making saline bridges with aspartate or glutamates residues that had to be broken to initiate the movement. In more recent versions, the charges are in water crevices in both the closed and open positions, decreasing the amount of translation required of the S4 segment.

The second class is that of the transporter model. In the closed position, the charges are in a water crevice connected to the intracellular solution, and in the open position they are in another water crevice connected to the extracellular solution. The translocation of the charges is achieved by a tilt and rotation of the S4 segment with little or no translation. In this case, the field is concentrated in a very small region that changes from around the first charge in the closed state to the fourth charge in the open state.

The third class is the paddle model introduced by the MacKinnon group (Jiang et al., 2003b), where the S4 segment is located in the periphery of the channel and the charges are embedded in the bilayer. The S4 segment makes a large translation such that the most extracellular charge goes from exposed to the extracellular medium in the open state to being completely buried in the bilayer in the closed state.

The helical screw and transporter models are similar but they differ dramatically from the paddle model in that the gating charges in the paddle model are embedded in the bilayer, while in the helical screw and transporter models the charges are surrounded by water or anions, or making salt bridges. In contrast with the transporter model, the helical screw model has in common with paddle models the large translation of the S4 segment.

At any rate, the results emerging from the described intense investigation on voltage-gated ion channels represent a vivid example of the role that charges and electric fields can have in key biological phenomena. In the next section, we will start considering another class of biological phenomena and reactions involving charge movement: the electron transfer.

3.7 Biological reactions and phenomena involving the transfer of electrons

Several among the most relevant biological phenomena are composed of single or chain reactions that involve the transfer of electrons along cascades of free energy. The best-known cases, but definitely not the only ones, are respiration and photosynthesis, the two mechanisms by which living beings store energy in the form of readily available chemical bonds in adenosine 5′-triphosphate (ATP).

We will see in the next sections some details of the electron transfer reactions involved in both oxidative phosphorylation (respiration) and photophosphorylation (photosynthesis). Generally speaking, electron-transfer reactions involve redox couples, that is, molecules that can exchange electrons by changing their oxidation state. A large number of redox reactions are involved in biological functions other than the aforementioned two, for instance, catalytic processes carried out by metalloenzymes, and thiol-disulfide exchange reactions. We will also consider these reactions in the next sections. Before entering into the details of specific, biologically relevant reactions, however, it is useful to review in general the mechanisms of biological electron transfer.

3.7.1 Biological electron transfer

At the basis of the marked interest in redox metalloproteins in bioelectronic applications there is their unique ability to perform electron-transfer reactions, i.e., to shuttle electrons between molecular partners as a crucial step in a large pool of different biological events (see section 4.1).

We have already mentioned that electron-transfer reactions are characteristic features of a variety of fundamental biological processes that include energy metabolism (photosynthesis, respiration, nitrogen fixation), hormone biosynthesis and xenobiotic detoxification. In most of the proteins involved in these processes, the active site is composed of a metal center, although organic cofactors (e.g., quinones, flavins, etc.) may also accomplish this function; see section 4.2. Irrespective of the specific features of each electron-transfer reaction, the basic principles that rule the electron-transfer rate are common to any such reaction. In order to understand the basis of the electron-transfer phenomenon and to define many of the mechanistic issues related to biological electron-transfer reactions, it is useful to consider the impulse from pioneering inorganic chemistry studies to the current biological research activities.

The genesis of the current approach to biological electron-transfer reactions dates back to the late 1940s, when coordination compounds with radio-labeled transition metals were used to study inorganic electron-transfer reactions (Marcus 1996). These initial studies enabled for the first time the determination of the rate constants for the transfer of an electron from the reduced to the oxidized state of a coordination compound (self-exchange rate constant). The availability of reliable experimental information for self-exchange reactions combined with the chemical simplicity of electron-transfer reactions (no bonds are formed or destroyed and the free energy of the products is identical to that of the reactants) attracted the attention of both experimentalists and theoreticians.

One can consider two different general types of inorganic electron transfer. The simplest one is that called outer-sphere electron transfer, in which the coordination shells of the reactants remain intact. It involves three steps:

$$A + D \leftrightarrow A\|D \tag{3.11}$$

$$A\|D \rightarrow A^-\|D^+ \tag{3.12}$$

$$A^-\|D^+ \leftrightarrow A^- + D^+ \tag{3.13}$$

The first corresponds to the formation of the so-called precursor complex (A||D), in which an electron acceptor (A) and a donor (D) interact through the ligands coordinated to the central metal atom.

Following the formation of a precursor complex, an electron is transferred from the donor to the acceptor, with the rate constant k_{et}, to form the successor complex ($A^-\|D^+$). The reaction is completed with the third step, the dissociation of the successor complex to the reaction products. Here, the rate of intramolecular electron transfer is influenced only by factors affecting k_{et}, whereas the intermolecular reactions are also a function of factors influencing the mutual interaction of the reactants (diffusion enabling the formation of the precursor complex, electrostatic interaction between reactants).

A different scenario applies for inner-sphere electron-transfer reactions. In this case, a reaction intermediate characterized by an inner coordination sphere of donor and acceptor sharing a common ligand of a metal ion is transiently formed. Generally speaking, coordination complexes that are relatively inert to ligand substitution follow outer-sphere electron-transfer mechanisms, whereas those complexes in which coordinating ligands are more labile are also more prone to undergo inner-sphere reactions. The requirement for formation of a ligand-bridged intermediate prior to electron transfer indicates that the energetic barrier for the formation of the transition state is significantly greater for the inner-sphere electron-transfer case. This difference in activation barriers between the two mechanisms reflects the Franck–Condon principle, which states that nuclear rearrangements are much slower than electronic ones.

The relative simplicity of the outer-sphere electron transfer reaction promoted the development of theoretical efforts towards the prediction of rate constants. In what follows, we will recall only the basic ideas, referring the interested reader to one of the numerous works reviewing the topic (Marcus, 1996; Marcus & Sutin, 1985; Devault, 1984; Moser & Dutton, 1996).

Let us start by considering the reaction described in equation (3.12) in the framework of the transition state theory (Eyring, 1935; Evans & Polanyi, 1935). In Figure 3.3 the potential energies of the precursor complex (A||D) and the successor one ($A^-||D^+$) are depicted as a function of the nuclear coordinate, representative of the nuclear motion. The Franck–Condon principle states that electronic transitions take place in times that are much shorter than those required for nuclei to move significantly. As a consequence, electron transitions take place at the crossing point R_{cp}. Mutual proximity of the acceptor and donor centers within the A||D complex causes electronic interaction between the centers that yields mixing and splitting of the two potential energy surfaces in the region of their intersection (the crossing point R_{cp}). The greater this electronic coupling (H_{AD}), the larger is the separation between the upper and the lower curves of Figure 3.3, and the higher is the probability that the reactants will proceed from the A||D curve to the $A^-||D^+$ one at R_{cp}. Reactions characterized by a strong electronic coupling are called "adiabatic"; in this case, the probability that the activated complex proceeds to products is nearly unity. For complexes in which this coupling is poor (as in the case where distances separating centers are large), the separation between upper and lower curves is small and there is an increased probability that reactants do not progress to the product

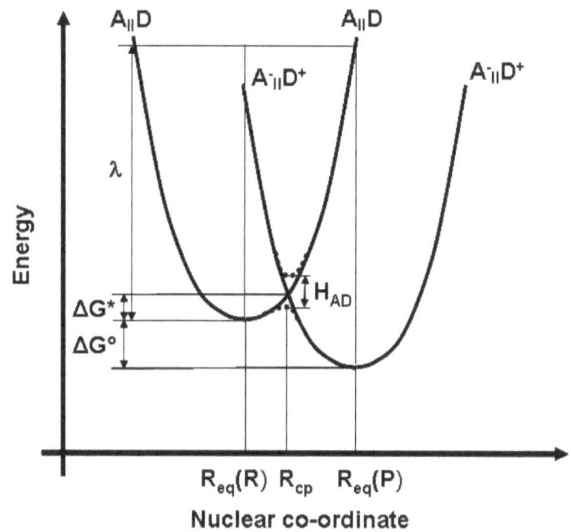

FIGURE 3.3

Energy diagram for an electron-transfer process between A and D. The potential energy of the nuclei for both the reactant and the products is represented by the two parabolas ($A_{||}D$ and $A^{-}{}_{||}D^{+}$) with equilibrium configuration $R_{eq}(R)$ and $R_{eq}(P)$. The passage from one parabola to the other represents the electron transfer process that occurs in the proximity of the crossing point R_{cp}. The activation energy for the reaction is ΔG^{*}, whereas the thermodynamic driving force is ΔG°. In the case of strong interaction between the donor and the acceptor, at the crossing point the two parabolas split, giving two curves separated by H_{AD}. This is referred to as an adiabatic transition with a smooth transition from the reactants to the products.

curve at the crossing point. These types of reactions are referred to as "non-adiabatic". The active sites of redox metalloproteins are quite insulating in nature and their metal ions are usually separated by relatively large distances. As a result, transfer of electrons in proteins is usually non-adiabatic, and mechanistic analysis of such reactions involves quantum-mechanical electron tunneling.

In order to compare experimental electron-transfer rate constants for non-adiabatic reactions with theoretical predictions, one needs to consider the following contributions:

- The magnitude of the electronic coupling between donor and acceptor as described by the matrix element H_{AD} (the separation between the A||D and the $A^{-}||D^{+}$ curve at the crossing point);
- The contribution of the Franck–Condon factor (FC).

All these aspects are accounted for by the fundamental equation of non-adiabatic electron transfer theory, according to Fermi's golden rule:

$$k_{et} = (2\pi\hbar)|H_{DA}|^2 FC \tag{3.14}$$

Equation (3.14) suggests that it is possible to decouple electronic and nuclear contributions in the electron-transfer rate. This is due to the fact that electron transfer occurs in a time too short for nuclei to modify their positions.

The magnitude of the matrix element H_{AD} depends on the nature of the donor and acceptor centers and the distance between them, and can be described by the Gamow relationship:

$$|H_{DA}|^2 = |H_M|^2 e^{-\beta R} \tag{3.15}$$

where R is the distance separating the centers, β is the decay length of electronic coupling with R, and H_M is the value of the matrix element for maximum coupling.

Marcus (see section 2.4.1) suggested that the FC factor in Equation (3.14) depends on:

- the energy λ associated with the rearrangement of the atomic nuclei of the reactants in their configuration in the products;
- the thermodynamic driving force for the reaction, $-\Delta G_0$, which is derived from the difference in the midpoint redox potential of donor and acceptor.

According to Marcus, the classic expression for FC factor is

$$FC = (4\pi\lambda k_B T)^{1/2} \exp\left[-(\lambda + \Delta G_0)^2/4\lambda k_B T\right] \tag{3.16}$$

where λ is the reorganization energy, k_B is the Boltzmann constant and $-\Delta G_0$ is the thermodynamic driving force of the reaction. $\ln(k_{et})$ varies parabolically with $-\Delta G_0$ such that k_{et} increases with the driving force until $-\Delta G_0 = \lambda$. As the driving force increases beyond this point, k_{et} decreases (if λ is constant). This critical prediction configures the so-called Marcus inverted region.

Therefore, the two dominant factors that contribute to shaping the electron-transfer rate constant are the distance between the donor and acceptor and the thermodynamic driving force of the reaction. A consequence of these considerations, first noted by Marcus, is that the cross reaction rate constant (k_{12}) for electron transfer between two substitutionally inert complexes (i.e., λ is nearly constant) is a function of the self-exchange rate constants for each of the reactants (k_{11} and k_{22}) and of the equilibrium constant of the reaction K_{12}, which is related to the difference in their reduction potentials (i.e., the thermodynamic driving force of the reaction):

$$k_{12} = (k_{11}k_{22}K_{12}f)^{1/2} \tag{3.17}$$

where

$$\ln f = (\ln K_{12})^2/4\ln(k_{11}k_{22}/z^2) \tag{3.18}$$

and z is the collisional frequency between neutral molecules in aqueous solution.

A large amount of experimental results provided compelling evidence for the validity of this theory.

Electron-transfer proteins, whose active sites are metal centers, have many features in common with simple coordination complexes. Metalloproteins that function solely in electron-transfer reactions possess metal centers that undergo minimal changes upon varying their oxidation state. On the other hand, metalloproteins that perform electron-transfer reactions as a part of catalytic cycles can exhibit more extensive structural rearrangements upon changing their oxidation state. In general, the location of the metal center within the protein and, hence, its accessibility by a redox partner allows a treatment of the electron-transfer behavior of these molecules according to the Marcus theory. Particularly, the electron-transfer rate constants measured for numerous redox metalloproteins revealed that fast, long-range electron transfer takes place at distances of up to 14 Å, involving inverse electron-transfer decay lengths of up to about 1 $Å^{-1}$.

At any rate, as we have already noted at the end of section 2.4.1, the Marcus theory represents the starting point for any further refinement, and still configures the most suitable framework for the interpretation of data involving the transport of electrons by proteins.

3.7.2 The photosynthetic reaction centers

The paramount importance that photosynthetic reaction centers (RCs) possess in biophysics stems not only from their relevance in the early stages of the most widespread approach that nature uses on the Earth to convert solar radiation in chemical energy, i.e., photosynthesis, but also from the fact that these molecules were the first transmembrane proteins for which a detailed 3D structure was established by X-ray crystallography and Fourier synthesis. Indeed, in 1982 Helmut Michel reported for the first time the X-ray diffraction pattern of a crystal of RCs from *Rhodopseudomonas viridis*, a photosynthetic purple bacterium. The corresponding structure was subsequently (1984) resolved at 3 Å by Michel, Deisenhofer and Huber (Deisenhofer et al., 1984), who were awarded the 1988 Nobel Prize in Chemistry for their remarkable achievement.

A photosynthetic RC converts the energy of captured sunlight into electrical and chemical energy during the first steps of photosynthesis by pumping protons from one side of a membrane to the other.

Molecular excitations, either originating directly from sunlight or transferred as excitation energy via light-harvesting antenna systems (vide infra) to the RC, give rise to electron-transfer reactions inside an RC along a series of protein-bound co-factors. These co-factors are light-absorbing pigments such as chlorophyll and pheophytin, as well as quinones. The energy of an incident photon is used to promote an electron to a higher molecular energy level of a pigment. The free energy created is then used to reduce a chain of nearby electron acceptors, which have subsequently higher redox potentials. These electron-transfer steps are the initial phases of a number of energy-conversion reactions, ultimately resulting in the production of chemical energy during photosynthesis.

RCs are present in all green plants, algae and many bacteria. Although these species are separated by billions of years of evolution, the RCs are characterized by a large degree of homology for all photosynthetic species. In contrast, a large variety in light-harvesting complexes exists between the photosynthetic species. Green plants and algae have two different types of RCs that are part of larger supercomplexes known as *photosystem I* (P700) and *photosystem II* (P680). These supercomplexes are large, involving multiple light-harvesting complexes. The RC found in *Rhodopseudomonas* bacteria is currently best understood not only because it was the first whose structure was unraveled, but also because it possesses fewer polypeptide chains than RCs in green plants.

The interior of the *Rhodopseudomonas* bacterium contains a large number of photosynthetic vesicles that are hollow, bilayer-contained, spheres. The photosynthetic RCs are embedded in the membrane of these vesicles, spanning the bilayer thickness from the periplasmic side (inner part of the vesicle) to the cytoplasmic side (external). Each RC is surrounded by about 100 smaller membrane proteins, the antenna pigment proteins. Each of these proteins contains several bound chlorophyll molecules that absorb photons over a large area, funneling them to the RC. Taking advantage of this arrangement, an RC can utilize 300 times more photons than those directly impinging on the special pair of (bacterio)chloropyll that is located at the heart of an RC.

A bacterial RC is made up of four polypeptide chains, three of which are called L, M and H for their respective putative molecular weights as deduced (erroneously) from SDS-PAGE electrophoretic mobility. Indeed, the number of amino acids in the sequences is 258 for H, 273 for L and 323 for M.

The L and M subunits have about 25% sequence homology, showing an evolutionary relationship. However, the sequence of the H subunit is completely different. The fourth subunit of the RC is a 336 amino acid cytochrome, lacking sequence similarity with any other known cytochrome. Furthermore, RC contains a number of pigments, namely, four bacteriochloropyll molecules, two of which form the so-called "special pair", one Fe atom, a carotenoid, two ubiquinone molecules and two bacteriopheophytins, which are basically apo-chlorophylls (chlorophylls lacking the central Mg^{2+} ion). The cytochrome subunit, instead, has four heme groups.

An RC contains 11 transmembrane helices made up of hydrophobic residues that anchor it firmly to the membrane. Five of them come from L and five from M subunits, whereas only one comes from the H subunit that is located at the cytoplasmic side of the center. This unique helix is at the C-terminus of the polypeptide chain and ends in the periplasmic space, close to the cytochrome location. The loop regions connecting the various α-helices are quite flat and hydrophilic, suitable to contact the cytochrome and the H subunit with the transmembrane part of the molecule.

Two helices from the L and two from the M subunit form a four-helix bundle that makes up the core of the molecule. A histidine from each of the four helices binds the Fe atom close to the cytoplasmic side. This Fe atom probably has a structure-stabilizing role, since it does not appear to be involved in the electron-transfer functionality. All the 11 transmembrane helices are tilted 20°–25° from the perpendicular to the membrane plane in a similar fashion to that found in bacteriorhodopsin and other transmembrane proteins.

The similar L and M subunits define a pseudo-twofold symmetry axis that runs between the pair of helices forming the protein core. The photosynthetic pigments are bound to these subunits and are also related by the same twofold symmetry axis. Thus, the pigments are arranged in a configuration that gives rise to two symmetric, possible electron-transfer pathways.

The special pair has the function of initiating the electron-transfer process. The two bacteriochlorophyll molecules are bound in a hydrophobic pocket in the vicinity of the symmetry axis, close to the periplasmic side of the membrane. Two pyrrol rings (one from each chlorophyll) are stacked together, 3 Å apart, and between them the symmetry axis passes. The other pigments are symmetrically arranged on each side of the symmetry axis. Another two bacteriochlorophyll molecules, the "accessory" ones, make hydrophobic contact with both the molecules of the special pair on the one hand and with the pheophytines on the other. In the functional RC there are also two quinones. One of them, Q_B, the final acceptor of the intramolecular ET chain, is only loosely bound and can be lost during protein purification. Although the reasons for this difference in quinone binding force are not clear, they are likely to reflect a functional asymmetry of the RC as a whole. Q_A has its polar headgroup outside the membrane and its hydrophobic tail bound to pheophytin and to the hydrophobic side chains of the L subunit transmembrane helices.

In photosynthesis, energy from light is converted to electrical energy by a charge separation process that leaves a positively charged donor and a negatively charged acceptor. Often, however, the electron-transfer back reaction, from acceptor to donor, is as fast as the forward one and the absorbed energy is lost, usually under the form of fluorescent radiation. The special arrangement of RCs, in both bacteria and green plants, allows for a very fast forward reaction and a slow back reaction, resulting in a stable charge separation upon light energy absorption. It is exactly this separation that stores energy and represents the basic primary process of photosynthesis.

In bacterial RCs, the process starts when light is absorbed by the special pair of BChl molecules located near the periplasmic side of the membrane. This pair of chlorophyll molecules absorbs photons between 870 nm and 960 nm, depending on the species and, thus, is called P870 (for the species *Rhodobacter sphaeroides*) or P960 (for *Rhodopseudomonas viridis*), where P stands for "pigment". Once P absorbs a photon, it releases an electron, which is transferred through another molecule of Bchl to the BPh in the L subunit. This initial charge separation yields a positive charge on P and a negative charge on the BPh. This process takes place in 10 picoseconds (10^{-11} seconds) (Devault 1984).

The charges on the special pair$^+$ and the BPh^- could undergo charge recombination in this state. This would waste the high-energy electron and convert the absorbed light energy into heat or fluorescent radiation. Several factors of the RC structure serve to prevent this. First, the transfer of an electron from BPh^- to $P960^+$ is relatively slow compared to two other redox reactions in the RC. The faster reactions involve the transfer of an electron from BPh^- (BPh^- is oxidized to BPh) to the electron acceptor quinone (Q_A) (taking place in about 200 ps), and the transfer of an electron to $P960^+$ ($P960^+$ is reduced to P960) from a heme in the cytochrome subunit above the RC.

The high-energy electron that resides on the tightly bound quinone molecule Q_A is transferred to the loosely bound quinone molecule Q_B in some 100 μs. This molecule is loosely associated with the protein and is fairly easy to detach. Two high-energy electrons are required to fully reduce Q_B to QH_2 (see section 4.3.2), taking up two protons from the cytoplasm in the process. The reduced quinone QH_2 diffuses through the membrane to another protein complex (cytochrome bc_1 complex), where it is oxidized. In the process the reducing power of the QH_2 is used to pump protons across the membrane to the periplasmic space. The electrons from the cytochrome bc_1 complex are then transferred through a soluble cytochrome c intermediate, called cytochrome c_2, in the periplasm to the cytochrome subunit. Thus, the flow of electrons in this system is cyclical. Although there are two such pigment arrangements, one at each side of the pseudo-twofold symmetry axis, only the pathway passing through the L subunit is used in RC charge separation.

The overall charge separation process is some eight orders of magnitude faster than the recombination reaction. This large difference allows RCs to have a remarkable *quantum efficiency* (number of charges separated per absorbed photons), between 98 and 100%, the energy yield (ratio between the energy stored by charge separation and photon energy) being about 50%.

Should the special pair of bacteriochlorophylls be the only photon acceptor in the bacterial membrane, only a small fraction of the incoming light would be absorbed by the RCs and converted into chemical energy. In all the photosynthetic organisms, however, a system of light-harvesting complexes is present that surrounds the RCs and contributes to increase the photon absorption cross-section by the membrane. Detailed structural information is available for the arrangement of these light-absorbing pigments around RCs in photosynthetic bacteria. The pigments are bound to small hydrophobic proteins embedded in the membrane of photosynthetic vesicles and are assembled in two different types of multimeric complexes called LH1 and LH2. Surprisingly, plants have apparently evolved completely different systems for light-harvesting with respect to photosynthetic bacteria.

Sticking to the light-harvesting assemblies of photosynthetic bacteria, the structure of LH2 has been elucidated in a number of cases, among which the first has been that of *Rhodopseudomonas acidophila*. It is a ring made of nine identical units, each containing two polypeptide chains of 53 and 41 residues, called α and β respectively, shaped in the form of alpha helices that span the membrane thickness. These two chains bind a total of three chlorophyll molecules and two carotenoids. The nine polypeptide units form a hollow cylinder where α chains occupy the internal and β chains the external wall cylinder. The hole inside this structure is occupied by membrane lipids.

The difference in the local environment of the three chlorophyll molecules endows them with different spectroscopic characteristics, namely they absorb light in a wavelength range wider than they would if they experienced the same environment. Indeed, one chlorophyll absorbs at 800 nm, whereas the other two, located near the periplasmic side of the membrane and oriented perpendicularly to the plane of the membrane, absorb at 850 nm.

LH2 from other bacteria can show differences, mainly related to the number of units making up the ring (e.g., eight, rather than nine in the case of *Rodospirillum molischianum*), but as far as the α and β polypeptide chains are concerned, a remarkable sequence similarity among different bacteria allows one to perform structural predictions.

This similarity also allows one to build a model for the other, core or inner antenna complex, LH1, closely associated with the reaction center in purple bacteria. Indeed LH1 surrounds the reaction center whereas LH2 is located around LH1 and is therefore more peripheral. Also LH1 is made of α and β chains with a certain sequence similarity to those of LH2. As a result, the ring-shaped disposition of chlorophyll molecules in the periplasmic ring of LH2 is preserved in LH1 but not the other chlorophyll (that absorbing at 800 nm), which in LH2 is bound by a sequence stretch that is absent in LH1. Indeed, the electron crystallography maps of LH1 (at 8.5 Å resolution) and LH2 show a remarkable similarity. However, LH1 contains 16 rather than 9 units and has a hollow structure 68 Å in diameter. The number of chlorophyll molecules in LH1 sums up presumably to 32 (16×2) and these appear to be in an even more hydrophobic environment given their absorption maximum shifted to 875 nm. Furthermore, the central hole of LH1 is occupied by a reaction center, the molecule where photons absorbed by antenna complexes are funneled.

In view of these structural results, complementing spectroscopic evidence and recent atomic force microscopy data of the purple membranes in physiological-like conditions (Scheuring et al., 2005), it is possible to figure out how the bacterial photosynthetic apparatus operates so efficiently.

As we have said, each reaction center sits in the central hole of an LH1 and is surrounded by 8 to 10 LH2 complexes. These numbers provide a figure of about 300 light-capturing chlorophyll molecules per reaction center.

LH2 absorbs radiation at a relatively shorter wavelength than LH1 and, following a downhill path, delivers it to LH1, which, in turn, passes it to the reaction center. Spectroscopic studies have shown that a photon absorbed by the 800 nm chlorophylls in LH2 is rapidly transferred to the 850 nm chlorophyll periplasmic ring in the same complex. The energy absorbed by one of these chlorophylls spreads to the others on a time scale of 0.2–0.3 ps, being readily delocalized all around the molecule. This fact also enables photon transfer to adjacent complexes, until they end up at the reaction center.

It is indeed worth noting that the primary photon acceptor–electron donor in the reaction center (the special pair of bacteriochloropylls) is located at the same level in the membrane (about 10 Å from the periplasmic side) as the 850 nm chlorophyll in LH2 and the 850 nm one in LH1. All of them have an orientation that enables an optimized energy transfer, making the photosynthetic apparatus extremely efficient in light energy to electron conversion (50% efficiency, to be compared with that of solid-state solar cells, which is around 20%).

Achieving a detailed understanding of this sophisticated mechanism may serve as an inspiration for the innovative design of synthetic, environmental friendly, solar cells operating under low light intensities.

3.7.3 Electron-transport chain in mitochondria and chloroplasts

Mitochondria

Most eukaryotic cells possess mitochondria, which produce ATP from products of the citric acid cycle, fatty acid and amino acid oxidation. At the mitochondrial inner membrane, electrons from NADH and succinate pass through the electron-transport chain to oxygen, which is reduced to water. The electron-transport chain includes an enzymatic series of electron donors and acceptors. Each electron donor passes electrons to a more electronegative acceptor, which in turn donates these electrons to another acceptor, a process that continues down the series until electrons are passed to oxygen, the most electronegative and terminal electron acceptor in the chain. Passage of electrons between donor and acceptor releases energy, which is used to generate a proton gradient across the mitochondrial membrane by actively "pumping" protons into the intermembrane space, producing a thermodynamic state that has the potential to do work. The entire process is called oxidative phosphorylation. During this process ADP is phosphorylated to ATP using the energy of hydrogen oxidation in many steps. A small percentage of electrons do not complete the whole series and instead directly leak to oxygen, causing the formation of the free-radical superoxide, a highly reactive molecular species that contributes to oxidative stress and is implicated in a number of diseases and aging.

Energy obtained through the transfer of electrons down the ETC is used to pump protons from the mitochondrial matrix into the intermembrane space, creating an electrochemical proton gradient across the mitochondrial inner membrane. This electrochemical proton gradient allows ATP synthase (ATP-ase) to use the flow of H^+ through the enzyme back into the matrix to generate ATP from adenosine diphosphate (ADP) and inorganic phosphate. Complex I (NADH coenzyme Q reductase) accepts electrons from the Krebs cycle electron-carrier nicotinamide adenine dinucleotide (NADH) and passes them to coenzyme UQ (ubiquinone), which also receives electrons from complex II (succinate dehydrogenase). UQ passes electrons to complex III (cytochrome bc_1 complex), which passes them to cytochrome *c* (cyt *c*). Cyt *c* passes electrons to complex IV (cytochrome *c* oxidase), which uses the electrons and protons to reduce molecular oxygen to water.

Four membrane-bound complexes have been identified in mitochondria. Each is an extremely complex transmembrane structure that is embedded in the inner membrane. Three of them are proton pumps. The structures are electrically connected by lipid-soluble electron carriers and water-soluble electron carriers.

Let us see in more detail the role of the various complexes. In complex I two electrons are removed from NADH and transferred to a lipid-soluble carrier, ubiquinone. The reduced product, ubiquinol (QH_2), freely diffuses within the membrane, and complex I translocates four protons across the membrane, thus producing a proton gradient. Complex I is one of the main sites at which premature electron leakage to oxygen occurs, thus being one of the main sites of production of superoxide (Nelson & Cox 2008).

The pathway of electrons occurs as follows: NADH is oxidized to NAD^+, by reducing flavin mononucleotide to $FMNH_2$ in one two-electron step. $FMNH_2$ is then oxidized in two one-electron steps, through a semiquinone intermediate. Each electron thus transfers from the $FMNH_2$ to an Fe-S cluster, from the Fe-S cluster to ubiquinone. Transfer of the first electron results in the free-radical (semiquinone) form of Q, and transfer of the second electron reduces the semiquinone form to the ubiquinol form, QH_2. During this process, four protons are translocated from the mitochondrial matrix to the intermembrane space (Nelson & Cox 2008).

In complex II additional electrons are delivered into the quinone pool originating from succinate and transferred (via FAD) to Q. Complex II consists of four protein subunits: SDHA, SDHB, SDHC and SDHD. Other electron donors (e.g., fatty acids and glycerol 3-phosphate) also direct electrons into Q (via FAD). Complex II is a parallel electron-transport pathway to complex I but, unlike complex I, no protons are transported to the intermembrane space along this pathway. Therefore, the pathway through complex II contributes less energy to the overall electron-transport chain process.

In complex III the Q-cycle contributes to the proton gradient by an asymmetric absorption/release of protons. Two electrons are removed from QH_2 at the Q_o site and sequentially transferred to two molecules of cyt *c*, a water-soluble electron carrier located within the intermembrane space. The two other electrons sequentially pass across the protein to the Q_i site where the quinone part of ubiquinone is reduced to quinol. A proton gradient is formed by two quinol ($4H^+4e^-$) oxidations at the Q_o site to form one quinol ($2H^+2e^-$) at the Q_i site (in total six protons are translocated: two protons reduce quinone to quinol and four protons are released from two ubiquinol molecules).

When electron transfer is reduced (by a high membrane potential or respiratory inhibitors such as antimycin A), complex III may leak electrons to molecular oxygen, resulting in superoxide formation.

In complex IV four electrons are removed from four molecules of cyt *c* and transferred to molecular oxygen, producing two molecules of water. At the same time, four protons are removed from the mitochondrial matrix (although only two are translocated across the membrane), contributing to the proton gradient.

According to the chemiosmotic coupling hypothesis, proposed by Peter D. Mitchell, the electron-transport chain and oxidative phosphorylation are coupled by a proton gradient across the inner mitochondrial membrane. The efflux of protons from the mitochondrial matrix creates an electrochemical gradient (proton gradient). This gradient is used by the FOF1 ATP synthase complex to make ATP via oxidative phosphorylation. ATP synthase is sometimes described as complex V of the electron-transport chain. The FO component of ATP synthase acts as an ion channel that provides proton flux back into the mitochondrial matrix. This reflux releases free energy produced during the generation of the oxidized forms of the electron carriers (NAD^+ and Q). The free energy is used to drive ATP synthesis, catalyzed by the F1 component of the complex.

Coupling with oxidative phosphorylation is a key step for ATP production. However, in specific cases, uncoupling the two processes may be biologically useful. The uncoupling protein, thermogenin – present in the inner mitochondrial membrane of brown adipose tissue – provides for an alternative flow of protons back to the inner mitochondrial matrix. This alternative flow results in thermogenesis rather than ATP production. Synthetic uncouplers (e.g., 2,4-dinitrophenol) also exist.

Chloroplasts

Whereas in oxidative phosphorylation electrons are transferred from a low-energy electron donor (e.g., NADH) to an acceptor (e.g., O_2) through an electron-transport chain, in photophosphorylation, which takes place in chloroplasts, the energy of sunlight is used to create a high-energy electron donor and an electron acceptor. Electrons are then transferred from the donor to the acceptor through another electron-transport chain.

In the thylakoid membranes of choloroplasts two kinds of photophosphorylation take place: cyclic and non-cyclic photophosphorylation.

In cyclic electron flow, the electron begins in a pigment complex called photosystem I, passes from the primary acceptor to ferredoxin, then to cytochrome b_6f (a similar complex to that found in mitochondria), and then to plastocyanin before returning to chlorophyll. This transport chain produces a proton-motive force, pumping H^+ ions across the membrane; this produces a concentration gradient that can be used to power ATP synthase during chemiosmosis. This pathway is known as cyclic photophosphorylation, and produces neither O_2 nor NADPH. Unlike non-cyclic photophosphorylation, $NADP^+$ does not accept the electrons; they are instead sent back to photosystem I.

The other pathway, noncyclic photophosphorylation, is a two-stage process involving two different chlorophyll photosystems. First, a water molecule is broken down into $2H^+ + 1/2\ O_2 + 2e^-$ by a process of photolysis. The two electrons from the water molecule are kept in photosystem II, while the $2H^+$ and $1/2O_2$ are left out for further use. Then a photon is absorbed by chlorophyll pigments that surround the reaction center of the photosystem. The light excites the electrons of each pigment, causing a chain reaction that eventually transfers energy to the core of photosystem II, exciting the two electrons that are transferred to the primary electron acceptor, pheophytin. The deficit of electrons is replenished by taking electrons from another molecule of water. The electrons transfer from pheophytin to plastoquinone, which takes the $2e^-$ from pheophytin and two H^+ from the stroma, and forms PQH_2, which later is broken into PQ; the $2e^-$ is released to cyt b_6f complex and the two H^+ are released into the thylakoid lumen. The electrons then pass through cyt b_6 and cyt f. Then they are passed to plastocyanin, providing the energy for protons to be pumped into the thylakoid space. This generates a gradient, making H^+ ions flow back into the stroma of the chloroplast, thus providing the energy for the regeneration of ATP.

The photosystem II complex replaces its lost electrons from an external source; however, the two other electrons are not returned to photosystem II as they would be in the analogous cyclic pathway. Rather, the still-excited electrons are transferred to a photosystem I complex, which boosts their energy to a higher level using a second solar photon. The highly excited electrons are transferred to the acceptor molecule, but this time they are passed on to an enzyme called ferredoxin-$NADP^+$ reductase, which uses them to reduce $NADP^+$ to NADPH.

This consumes protons produced by the splitting of water, leading to a net production of $1/2O_2$, ATP and NADPH+H^+ with the consumption of solar photons and water.

The concentration of NADPH in the chloroplast may help regulate which pathway electrons take through the light reactions. When the chloroplast runs low on ATP for the Calvin cycle, NADPH will accumulate and the plant may shift from noncyclic to cyclic electron flow.

3.7.4 Thiol-disulfide exchange reactions

A further important example of phenomena involving electron transfer in biology is represented by the thiol-disulfide exchange reaction. Despite its general importance, we will focus here on this reaction also because it will represent a key tool in the proposed approach towards the control of gene expression level in photosynthetic bacteria (see Chapters 6, 7 and 8).

Thiols groups and disulfide bonds undergo a spontaneous chemical reaction, in which a thiol displaces one sulfur atom of a disulfide bond.

This type of reaction can be found in many biological phenomena, especially those involving the thiol groups of cysteine residues of proteins, thioredoxin, glutaredoxin, protein disulfide isomerase, glutathione, coenzyme A, dihydrolipoamide and lipoic acid. This thiol-disulfide exchange reaction possesses unique features in its ability to break and reform a strong, directional, covalent, disulfide bond in water at room temperature reversibly at a rapid rate that can be varied over many orders of magnitude by the effect of pH. From a technological standpoint, it is interesting to note that this reaction is one of the most specific chemical reactions in molecular biology, since thiol groups and disulfide bonds tend not to react with other groups on proteins, nucleic acids, etc., constituting a very powerful tool in analysis and synthesis.

The ionized form of the thiol group is the reactive species; therefore, below the pK_a of the thiol group, the reaction rate varies with pH. Typical biological thiol groups of cysteine residues have pK_a values around 9. The reaction can be quenched by acidification, making the thiolate anion irrelevant. Alternatively, such mixtures can be trapped irreversibly by rapidly reacting all available thiols with reagents such as iodoacetamide, iodoacetate, or *N*-ethylmaleimide. The thiol and disulfide species present in both equilibrium and kinetically trapped mixtures can be analyzed with analytical chemistry tools such as HPLC or NMR, under acidic conditions. The rate of a simple thiol-disulfide exchange reaction can be measured directly if the disulfide bonds reacting and being generated differ in their spectrophotometric responses. For example, cyclic disulfide bonds in 5- or 6-membered rings, such as that formed in dithiothreitol, absorb radiation in the interval 280–330 nm, whereas most linear disulfide bonds do not. Dynamic 1H-NMR lineshape analysis and spin-transfer methods have also been used to measure the rate constants of simple model reactions.

The three sulfur atoms involved in a redox reaction are generally labeled as the nucleophile (nuc), the central atom (c) and the leaving group (lg). The reaction takes place by the ionized thiolate anion nucleophile attacking one of the two sulfur atoms forming the disulfide bond, which becomes the central atom, optimally along the axis of the S-S bond. The transition state for the reaction is expected to be a linear arrangement of the three sulfur atoms, equally spaced, and with the negative charge of the thiolate anion spread asymmetrically, more on the terminal atoms than on the central one. The reaction rate constant depends upon the electron affinities of each of the three sulfur atoms, which are conveniently measured by their pK_a values when in the form of thiols. The second-order rate constant for the intermolecular reaction between model compounds can be predicted by their three pK_a values:

$$log\ k_{RS-} = 4.5 + 0.59pK_a^{nuc} - 0.40pK_a^{c} - 0.59pK_a^{lg} \tag{3.19}$$

$$log k_{obs} = 4.5 + 0.59pK_a^{nuc} - 0.40pK_a^{c} - 0.59pK_a^{lg} - \log\left(1 + 10^{pK^{nuc} - pH}\right) \tag{3.20}$$

Of course, it is not possible to measure just one thiol-disulfide exchange reaction, such as that of equation (3.19), because the thiolate anion will also react with the other sulfur atom of the disulfide bond, and the combined rates will be measured as any initial rate. As a consequence, the mixed disulfides formed will also react further with other thiol groups, yielding complex mixtures of disulfide and thiol species. The second-order rate constant for reaction of a typical, fully ionized thiol group with a typical cysteine disulfide bond is approximately 20 M^{-1} s^{-1} at 25 °C. The optimal pK_a for the attacking thiol group, i.e., that providing the largest rate, corresponds to the pH of the reaction. In the case of a thiol with a higher pK_a value, the ionized form will be less abundant; the thiol with a lower pK_a value is ionized, but its nucleophilicity is lower.

The above considerations apply equally to the reverse reaction, and the expected equilibrium constant can be estimated from equation (3.21) for the reaction in both directions. If the attacking and departing sulfur atoms have different pK_a values, the equilibrium constant is pH-dependent; it varies over the pH interval defined by the two thiol pK_a values. The equilibrium favors the thiol group with the lower pK_a value; just the opposite would have been predicted considering only the effect of ionization of the two thiols on mass action.

The reaction occurs more rapidly than predicted if there are positive charges near the disulfide bond, to attract the attacking thiolate anion, or if the disulfide bond is strained. For example, the strained disulfide bond of lipoic acid reacts considerably more rapidly than otherwise expected, which is accounted for by about 3.8 kcal/mol of conformational strain in the five-membered ring (the CSSC dihedral angle is only about 30° and not the favored 90°). Conversely, the reaction is inhibited by the presence of negative charges near the disulfide bond, if either the thiol or the disulfide group is buried and inaccessible, or if there are bulky substituents adjacent to the sulfur atoms. Electrostatic effects on the equilibria of thiol-disulfide exchange reactions are small in magnitude, but occur in the expected direction. Formation of a mixed disulfide with different charges close to the two moieties is favored, whereas that with equal charges is disfavored. Such effects are substantial only when the charged groups are on adjacent residues, and their magnitude can be decreased by the electrostatic screening action of high salt concentrations.

The thiol-disulfide exchange reaction is often used in molecular biology when protein disulfide bonds (PSS) are reduced by reagents such as dithiothreitol or β-mercaptoethanol. In this case, two sequential thiol-disulfide exchange reactions are necessary, proceeding through a mixed disulfide between the reagent (RSH) and the protein:

$$\mathrm{P_S^S} + 2\mathrm{RSH} \leftrightarrow \mathrm{P_{SH}^{SSR}} + \mathrm{RSH} \leftrightarrow \mathrm{P_{SH}^{SH}} + \mathrm{RSSR} \qquad (3.21)$$

Some proteins, such as thioredoxin, glutaredoxin and protein disulfide isomerase, participate in thiol-disulfide exchange reactions much more rapidly than expected from the pK_a values of their active-site thiol groups; equation (3.20). This is believed to be due to the tendency of these proteins to bind noncovalently the molecules that they react with, in a fashion similar to an enzyme binding its substrate, and thus stabilizing the transition state for the reaction, again like an enzyme. When present in small quantities, such proteins appear to catalyze thiol-disulfide exchange reactions between other thiol and disulfide moieties. They do this by reacting more rapidly with both the thiol and disulfide compounds, so they do not catalyze the direct reaction between them but instead provide a more rapid alternative reaction pathway.

Interestingly, thiol-disulfide exchange reactions are readily reversible, and the above reactions can be used to add disulfide bonds to a protein.

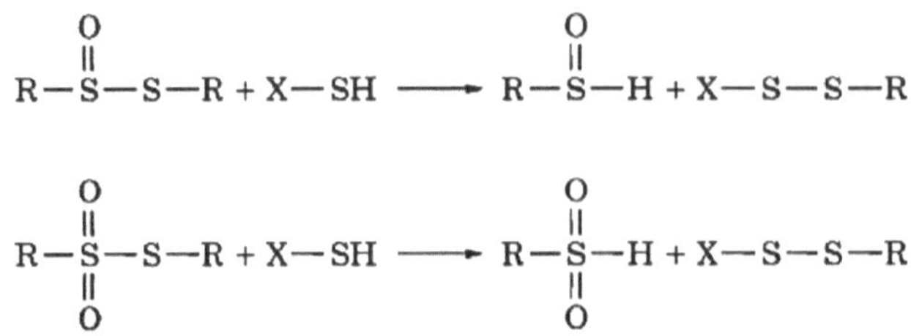

SCHEME 3.1

Thiol-disulfide exchange reactions in presence of a mono- (upper) and dioxide (lower) form of disulfide.

The thiol-disulfide exchange reaction can be simplified if a mono- or dioxide form of the disulfide is used. The oxidation increases the reactivity, but only the nonoxidized sulfur atom undergoes thiol-disulfide exchange, according to Scheme 3.1.

As a consequence, the reaction stops at this stage and becomes stoichiometric.

3.8 References

Agnew, W. S., Levinson, S. R., Brabson, J. S., & Raftery, M. A. (1978). Purification of the tetrodotoxin-binding component associated with the voltage-sensitive sodium channel from electrophorus electricus electroplax membranes. *Proc. Natl. Acad. Sci. USA, 75*, 2602–2610.

Arai, S., Chatake, T., Ohhara, T., Kurihara, K., Tanaka, I., Suzuki, N., Fujimoto, Z., Mizuno, H., & Niimura, N. (2005). Complicated water orientations in the minor groove of the B-DNA decamer d(CCATTAATGG)2 observed by neutron diffraction measurements. *Nucleic Acids Res., 33*, 3017–3024.

Bandyopadhyay, S., Chakraborty, S., & Bagchi, B. (2005). Secondary structure sensitivity of hydrogen bond lifetime dynamics in the protein hydration layer. *J. Am. Chem. Soc., 127*, 16660–16667.

Ben-Naim, A. (2002). Molecular recognition — viewed through the eyes of the solvent. *Biophys. Chem., 101–102*, 309–319.

Berg, J. M., Stryer, L., & Tymoczko, J. L. (2002). *Biochemistry* (5th ed.). New York: Freeman.

Brovchenko, I., Krukau, A., Smolin, N., Oleinikova, A., Geiger, A., & Winter, R. (2005). Thermal breaking of spanning water networks in the hydration shell of proteins. *J. Chem. Phys., 123*(224905), 1–10.

Chaplin, M. F. (2004). The importance of cell water. *Sci. Soc., 24*, 42–45.

Chaplin, M. (2006). Do we underestimate the importance of water in cell biology? *Nat. Rev. Mol. Cell Biol., 7*, 861–866.

Cheung, M. S., Garcia, A. E., & Onuchic, J. N. (2002). Protein folding mediated by solvation: water expulsion and formation of the hydrophobic core occur after the structural collapse. *Proc. Natl. Acad. Sci. USA, 99*, 685–690.

Ciccotti, G., Frenkel, D., & McDonald, I. R. (1987). *Simulation of liquids and solids.* Amsterdam: North Holland.

Coulson, C. A. (1961). *Valence* (2nd ed.). New York: Oxford University Press.

Cukierman, S. (2006). Et tu, Grotthuss! and other unfinished stories. *Biochim. Biophys. Acta Bioenergetics, 1757*, 876–885.

Dashevsky, V. G., & Sarkisov, G. N. (1974). Solvation and hydrophobic interaction of non-polar molecules in water in approximation of interatomic potentials: the Monte-Carlo method. *Molecular Physics, 27*, 1271–1290.

Deisenhofer, J., Epp, O., Miki, K., Huber, R., & Michel, H. (1984). X-ray structure analysis of a membrane protein complex. Electron density map at 3 Å resolution and a model of the chromophores of the photosynthetic reaction center from Rhodopseudomonas viridis. *J. Mol. Biol., 180*, 385–398.

Devault, D. (1984). *Quantum mechanical tunnelling in biological systems.* Cambridge: Cambridge University Press.

Driscoll, R. J., Youngquist, M. G., & Baldeschwieler, J. D. (1990). Atomic-scale imaging of DNA using scanning tunnelling microscopy. *Nature, 346*, 294–296.

Eisenmesser, E. Z., Millet, O., Labeikovsky, W., Korzhnev, D. M., Wolf-Watz, M., Bosco, D. A., Skalicky, J. J., Kay, L. E., & Kern, D. (2005). Intrinsic dynamics of an enzyme underlies catalysis. *Nature, 438*, 117–121.

Evans, M. G., & Polanyi, M. (1935). Some applications of the transition state method to the calculation of reaction velocities, especially in solution. *Trans. Faraday Soc., 31*, 875–894.

Eyring, H. (1935). The activated complex in chemical reactions. *J. Chem. Phys., 3*, 107–115.

Fernández, A., & Scheraga, H. A. (2003). Insufficiently dehydrated hydrogen bonds as determinants of protein interactions. *Proc. Natl. Acad. Sci. USA, 100*, 113–118.

Fuxreiter, M., Mezei, M., Simon, I., & Osman, R. (2005). Interfacial water as a hydration fingerprint in the noncognate complex of BamHI. *Biophys. J., 89*, 903–911.

Gray, H. B., & Winkler, J. R. (2003). Electron tunneling through proteins. *Q. Rev. Biophys., 36*, 341–372.

Gray, H. B., & Winkler, J. R. (2005). Long-range electron transfer. *Proc. Natl. Acad. Sci. USA, 102*, 3534–3539.

Guillot, B. (2002). A reappraisal of what we have learnt during three decades of computer simulations on water. *J. Mol. Liquids, 101*, 219–260.

Halle, B. (2004). Protein hydration dynamics in solution: a critical survey. *Phil. Trans. R. Soc. Lond. B., 359*, 1207–1224.

Hamill, O. P., Marty, A., Neher, E., Sackmann, B., & Sigworth, F. J. (1981). Improved patch-clamp techniques for high-resolution current recording from cells and cell-free membrane patches. *Pflugers Arch., 391*, 85–100.

Hasted, J. B. (1973). *Aqueous dielectrics.* London: Chapman & Hall.
Hille, B. (2001). *Ion channels of excitable membranes* (3rd ed.). Sunderland, MA: Sinauer Associates.
Hirschberg, B., Rovner, A., Lieberman, M., & Patlak, J. (1996). Transfer of twelve charges is needed to open skeletal muscle Na channels. *J. Gen. Physiol., 106*, 1053–1068.
Hobbs, P. V. (1974). *Ice physics.* Oxford: Clarendon Press.
Hodgkin, A. L., & Huxley, A. F. (1952). A quantitative description of membrane current and its application to conduction and excitation in nerve. *J. Physiol., 117*, 500–544.
Hollins, G. T. (1964). Configurational statistics and the dielectric constant of ice. *Proc. Phys. Soc., 84*, 1001.
Hoshi, T. (1995). Regulation of voltage dependence of the KAT1 channel by intracellular factors. *J. Gen. Physiol., 105*, 309–328.
Israelachvili, J. N. (2005). *Intermolecular and surface forces* (3rd ed.). New York: Academic Press.
Israelachvili, J. N., & Pashley, R. M. (1982). The hydrophobic interaction is long range, decaying exponentially with distance. *Nature, 300*, 341–342.
Jeffrey, G. A. (1997). *An introduction to hydrogen bonding (topics in physical chemistry).* New York: Oxford University Press.
Jiang, Y., Lee, A., Chen, J., Ruta, V., Cadene, M., Chait, B. T., & MacKinnon, R. (2003a). X-ray structure of a voltage-dependent K^+ channel. *Nature, 423*, 33–41.
Jiang, Y., Ruta, V., Chen, J., Lee, A., & MacKinnon, R. (2003b). The principle of gating charge movement in a voltage-dependent K^+ channel. *Nature, 423*, 42–48.
Joesten, M. D., & Schaad, L. J. (1974). *Hydrogen bonding*. New York: Dekker.
Kunst, M., & Warman, J. M. (1980). Proton mobility in ice. *Nature, 288*, 465–467.
Li, Z., & Lazaridis, T. (2005). The effect of water displacement on binding thermodynamics: concanavalin A. *J. Phys. Chem. B., 109*, 662–670.
Lin, J., Balabin, I. A., & Beratan, D. N. (2005). The nature of aqueous tunneling pathways between electron-transfer proteins. *Science, 310*, 1311–1313.
Luzar, A., Bratko, D., & Blum, L. J. (1987). Monte-Carlo simulation of hydrophobic interaction. *Chem. Phys., 86*, 2955–2959.
Mahoney, M. W., & Jorgensen, W. L. (2000). A five-site model for liquid water and the reproduction of the density anomaly by rigid, nonpolarizable potential functions. *J. Chem. Phys., 112*, 8910–8922.
Makarov, V., Pettitt, B. M., & Feig, M. (2002). Solvation and hydration of proteins and nucleic acids: a theoretical view of simulation and experiment. *Acc. Chem. Res., 35*, 376–384.
Marcelja, S., Mitchell, D. J., Ninham, B. W., & Sculley, M. J. (1977). Role of solvent structure in solution theory. *J. Chem. Soc. Faraday Trans., 2*, 630–648.
Marcus, R. A. (1996). Electron transfer reactions in chemistry and biology. Theory and experiments. In D. S. Bendall (Ed.), *Protein electron transfer.* Oxford: BIOS Scientific.

Marcus, R. A., & Sutin, N. (1985). Electron transfers in chemistry and biology. *Biochim. Biophys. Acta, 811*, 265–322.

Moser, C. C., & Dutton, P. L. (1996). Outline of theory of protein electron transfer. In D. S. Bendall (Ed.), *Protein electron transfer.* Oxford: BIOS Scientific.

Nakasako, M. (2004). Water–protein interactions from high resolution protein crystallography. *Phil. Trans. R. Soc. Lond. B., 359*, 1191–1206.

Nelson, D. L., & Cox, M. M. (2008). *Lehninger principles of biochemistry* (5th ed.). New York: W. H. Freeman.

Nicholson, D., & Parsonage, N. G. (1982). *Computer simulation and the statistical mechanics of adsorption.* New York: Academic Press.

Noceti, F., Baldelli, P., Wei, X., Qin, N., Toro, L., Birnbaumer, L., & Stefani, E. (1996). Effective gating charges per channel in voltage-dependent K^+ and Ca^{2+} channel. *J. Gen. Physiol., 108*, 143–155.

Noda, M., Shimizu, S., Tanabe, T., Takai, T., Kayano, T., Ikeda, T., Takahashi, H., Nakayama, H., Kanaoka, Y., & Minamino, N. (1984). Primary structure of *Electrophorus electricus* sodium channel deduced from cDNA sequence. *Nature, 312*, 121–127.

Pangali, C., Rao, M., & Berne, B. J. (1979). A Monte Carlo simulation of the hydrophobic interaction. *J. Chem. Phys., 71*, 2975–2981.

Parsegian, A. (1969). Energy of an ion crossing a low dielectric membrane: solutions to four relevant electrostatic problems. *Nature, 221*, 844–846.

Pauling, L. (1935). The structure and entropy of ice and of other crystals with some randomness of atomic arrangement. *J. Am. Chem. Soc., 57*, 2680–2684.

Pauling, L. (1960). *The nature of the chemical bond.* Ithaca, NY: Cornell University Press.

Ponce, A., Winkler, J. R., & Gray, H. B. (2000). Electron tunneling in water: oxidative quenching of electronically excited $Ru(tpy)_2^{2+}$ (tpy = 2,2':6,2''-terpyridine) by ferric ions in aqueous glasses at 77 K. *J. Am. Chem. Soc., 122*, 8187–8191.

Pratt, L. R., & Chandler, D. (1977). Theory of the hydrophobic effect. *J. Chem. Phys., 67*, 3683–3704.

Scheuring, S., Lévy, D., & Rigaud, J.-L. (2005). Watching the components of photosynthetic bacterial membranes and their in situ organisation by atomic force microscopy. *Biochim. Biophys. Acta Biomembranes, 1712*, 109–127.

Schuster, P., Zundel, G., & Sandorfy, C. (1976). The hydrogen bond. Amsterdam: North-Holland.

Seoh, S.-A., Sigg, D., Papazian, D. M., & Bezanilla, F. (1996). Voltage-sensing residues in the S2 and S4 segments of the Shaker K^+ channel. *Neuron, 16*, 1159–1167.

Smolin, N., Oleinikova, A., Brovchenko, I., Geiger, A., & Winter, R. (2005). Properties of spanning water networks at protein surfaces. *J. Phys. Chem. B., 109*, 10995–11005.

Tajkhorshid, E., Nollert, P., Jensen, M. O., Miercke, L. J. W., O'Connell, J., Stroud, R. M., & Schulten, K. (2002). Control of the selectivity of the aquaporin water channel family by global orientational tuning. *Science, 296*, 525–530.

Tempel, T. M., Papazian, D. M., Schwarz, T. L., Jan, Y. N., & Jan, L. Y. (1987). Sequence of a probable potassium channel component encoded at Shaker locus in drosophila. *Science, 237*, 770–775.

Tezcan, F. A., Crane, B. R., Winkler, J. R., & Gray, H. B. (2001). Electron Tunneling in Protein Crystals. *Proc. Natl. Acad. Sci. USA, 98*, 5002–5006.

van Amsterdam, I. M., Ubbink, M., Einsle, O., Messerschmidt, A., Merli, A., Cavazzini, D., Rossi, G. L., & Canters, G. W. (2002). Dramatic modulation of electron transfer in protein complexes by crosslinking. *Nat. Struct. Biol., 9*, 48–52.

Vandenberg, C. A., & Bezanilla, F. (1991). A sodium channel model of gating based on single channel, macroscopic ionic and gating currents in the squid giant axon. *Biophys. J., 60*, 1511–1533.

Vaught, A., Jing, T. W., & Lindsay, S. M. (1995). Non-exponential tunneling in water near an electrode. *Chem. Phys. Lett., 236*, 306–310.

Wenger, O. S., Leigh, B. S., Villahermosa, R. M., Gray, H. B., & Winkler, J. R. (2005). Electron tunneling through organic molecules in frozen glasses. *Science, 307*, 99–102.

CHAPTER

Applications of Electrochemistry to Redox Metalloproteins and Cofactors

4

CHAPTER OUTLINE

4.0 Redox metalloprotein and cofactor electrochemistry

In Chapters 2 and 3 we have seen that electrochemical techniques are valuable tools for retrieving information on both the kinetics and the thermodynamics of electron-transfer processes involving redox biomolecules. Along with these analytical aspects, electrochemistry can also be used to control redox reactions occurring at electrodes, where molecules, freely diffusing in solution or adsorbed on electrode surfaces, can exchange electrons. In the next sections we will focus our attention on two special kinds of

Biomolecular Electronics. http://dx.doi.org/10.1016/B978-1-4557-3142-8.00004-2

biomolecules: redox metalloproteins and redox cofactors, which are usually involved in complex electron transfer processes in many biological systems. These classes of biomolecules represent paradigmatic cases whose study can both help us by shedding light on the molecular mechanisms underlying biological electron transfer, and provide clues towards the exploitation of their properties for technological purposes. Potential applications encompass single-molecule biotransistors or biomolecular switches, pH-gated electron transfer, as well as the fabrication of hybrid biocatalytic electrodes that take advantage, for instance, of surface-immobilized redox enzymes. Here, we are particularly interested in direct electrochemistry, i.e., direct electron exchange between biomolecules and electrodes. As a matter of fact, some electrochemical approaches involving biomolecules make use of so-called redox mediators, which are a sort of buffer redox molecular stage, usually adsorbed on the electrodes' surface, that shuttle electrons between a metal substrate and a redox biomolecule. Rather, the examples reported in the next sections will concern either freely diffusing molecules interacting with metal electrodes or redox moieties physisorbed on metal surfaces, possibly through a bifunctional linker.

In the next four sections we will introduce, in a general way, the electrochemical behavior of both redox metalloproteins and some redox cofactors relevant to the aims of the present book, showing some of the most important results concerning both their electrochemical characterization in diffusion and after adsorption on the surface of a working electrode as submonolayers. Lastly, we will address their single-molecule behavior with the help of a special tool, the electrochemical scanning tunneling microscope, that has helped us gain much insight into the mechanisms by which electrons can flow through single redox molecules, exploiting their inherent redox levels.

4.1 Redox metalloproteins

The transfer of electrons by biological macromolecules represents the means by which a relevant number of complex biological functions are partially or fully accomplished (Gray & Ellis, 1994). Key phenomena such as photosynthesis, respiration, catalytic reactions, and many more, involve as single or multiple crucial steps the transfer of one ore more electrons between molecular partners along free energy cascades. In biological electron transfer, a peculiar role is played by a special class of proteins called redox metalloproteins. These belong to the larger ensemble of metalloproteins, large polypeptides containing one or more metal ions or clusters that represent a fraction between 25 and 30% of the entire proteome (Cowan, 1997).

Generally speaking, the set of biologically relevant metals (i.e., those important for biological functions) includes iron, magnesium, copper, nickel, zinc, manganese, molybdenum, cobalt, tungsten and vanadium (Holm et al., 1996). These metals and their ligands define prosthetic groups that are usually covalently bound to the polypeptide backbone (e.g., the heme group in cytochromes) by endogenous ligands brought about by amino acid side chains.

Redox or electron-transfer metalloproteins have one or more metal ions in their active site, whose oxidation state can change reversibly, forming the basis of their ability to exchange electrons with other biomolecular partners. The most relevant among these metalloproteins can be classified into three families: blue copper proteins (or cuprodoxins), cytochromes (containing heme groups) and iron-sulfur complexes; to these one can add the chlorophyll-based photosynthetic complexes (e.g., reaction centers, antenna complexes) that are redox metalloproteins where a photoionization event triggers an electron transfer cascade (Devault, 1984).

Given their peculiar functional activity and extreme biological relevance, redox metalloproteins have been the object of intense study by theoretical, computational and experimental approaches (Harrison, 1985). These range from classical and semi-classical (QM/MM) molecular dynamics, to various spectroscopies (UV-vis and IR absorption, fluorescence, Raman scattering, EXAFS, quasi-elastic and low-angle neutron scattering, NMR, EPR, XPS), diffractions (X-rays, electrons), pulsed radiolysis, electrochemistry, and, more recently, scanning probe microscopy (scanning tunneling microscopy (STM), electrochemically assisted STM, atomic force microscopy (AFM)), often assisted by protein engineering to help shed light on the fine structural and functional details of proteins by means of site-directed mutagenesis. The concerted scientific effort directed towards metalloprotein characterization has given rise to a deep understanding of the structure and function of several of these redox molecules and, in many cases, it has allowed us to shed light on the mechanisms by which these biomolecules shuttle electrons between molecular partners.

This aspect is, of course, very important in view of the possible technological exploitation of redox metalloprotein electron transfer. Indeed, the advent of nanoscience has opened up the possibility of attaining an unprecedented level of understanding of the electron-transfer mechanisms ruling the functional activity of redox metalloproteins, that of the single molecule. Indeed, at that level, one can get rid of the statistical average typical of "bulk" investigation methods (which tend to smear out the fine differences in the possible electron-transfer mechanisms; vide infra), enabling us attain a quantitative, mechanistic description of the observed phenomena and a related control.

It is worth noting that the efficiency with which redox metalloproteins perform the task of transferring electrons, along with their nanometer size and the self-assembly ability inherent to many biological molecules, makes them very interesting candidates for technological applications (Facci, 2002), namely for the implementation of bioelectronic devices.

In the remainder of this chapter we will see, as examples, the demonstration of a nanotransistor-like functionality of a particular blue copper protein, the azurin, and the use of engineered cytochrome *c* for assembling oriented submonolayers of functional redox proteins.

4.1.1 The azurin

Among electron-transfer metalloproteins, an important set is the so-called "blue copper proteins". It is composed of redox biomolecules bearing a single Cu ion in their active site that is characterized by a distorted coordination geometry that causes their typical intense blue color. Blue copper proteins, or cuprodoxins, can be found in both bacteria and plants, where they are usually involved in some of the early steps of respiration and photosynthesis (Adam, 1991). Their family is mainly composed of azurin, plastocyanin, amicyanin, rusticyanin and ceruloplasmin. In particular, azurin has been thoroughly investigated by electrochemically assisted scanning probe techniques because of a series of peculiar features.

Azurin from *Pseudomonas aeruginosa* is a water-soluble, relatively small protein (molecular mass 14,600) that is involved in the oxidative phosphorylation of the bacterium that expresses it. Its functional role is believed to be that of shuttling electrons between two molecular partners, cytochrome c_{551} and nitrite reductase, that act as azurin primary donor and acceptor, respectively (Bendall, 1996). From a redox chemistry standpoint, azurin's functional behavior depends on the reversible oxidation of Cu^{1+} to Cu^{2+}. Peculiar electronic properties are connected to the special structural features of its redox site. Indeed, it contains a copper ion ligated to five amino acids (two histidines and a cysteine, strongly bound to copper, and two weaker axial ligands: a methionine and a main chain carbonyl oxygen) (Nar et al., 1991). The resulting geometry is determined by a peculiar ligand-field symmetry that provides the site with unusual spectroscopic and electrochemical characteristics. Among them are an intense optical absorption band at

628 nm (associated with a S(Cys)–Cu bonding to antibonding transition), a small hyperfine splitting in the electron paramagnetic spectrum (Brill, 1977) and a surprisingly large standard potential (+116 mV vs SCE) (Chi et al., 1999) in comparison with the Cu(II/I) aqua redox couple (−89 mV vs SCE) (Lide, 1993). Its structure has a globular shape characterized mainly by a β-barrel conformation; Figure 4.1. The azurin active redox site is buried inside the molecule and appears located ≈ 0.8 nm below the globule surface; it also features an exposed S-S bridge that is formed by the thiol moieties of the Cys3 and Cys26 side chains.

Both its conformation and the presence of the above-mentioned disulfur bridge make azurin very appealing as a potential candidate for single-molecule-level investigation by scanning probe techniques.

Indeed, the presence of an exposed disulfide provides a very convenient anchoring point to chemisorb azurin at the surface of electronically soft metals such as Au, Ag, Pt, Cu, etc. by means of the formation of one or two Me-S

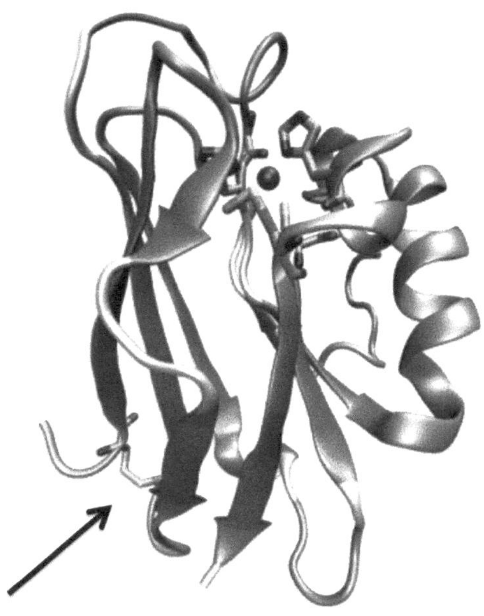

FIGURE 4.1

Schematic representation of the structure of azurin from *Pseudomonas aeruginosa*. Note the Cu ion represented as a dot in the upper part of the structure; the arrow points to the disulfide bridge formed by Cys3 and Cys26 thiol oxidation. Structural data from PDB file 1E5Y.

bonds. A relevant number of experimental techniques, including FTIR absorption spectroscopy in ATR configuration, quartz crystal microbalance (QCM), scanning force microscopy, XPS, reductive desorption and cyclic voltammetry, confirm that azurin chemisorbs on Au and that it retains its redox activity once arranged in a submonolayer (Schnyder et al., 2002). Finally, the aforementioned β-barrel conformation provides the protein with a remarkable resistance to the mechanical action of the scanning probe, particularly useful in the case of thoroughly repeated scans. This is a very desirable feature that unfortunately is not very common in biomolecules, which tend to be damaged by the action of a scanning tip, even in those cases where tip–sample interaction can be kept to a minimum (e.g., alternating contact mode AFM; Alessandrini & Facci, 2005).

4.1.2 Cytochrome *c*

Cytochrome *c* (cyt *c*), like azurin, is also a small, monomeric water-soluble protein. At variance with azurin, however, it belongs to the heme-protein family. Physiologically, it is located in the intermediate mitochondrial space where it performs the task of shuttling electrons between ubiquinol/cytochrome *c* oxidoreductase (QCR, cytochrome bc_1 complex) and cytochrome *c* oxidase (Lange & Hunte, 2002; Saraste, 1999).

Cyt *c* is one of the most-studied redox proteins because of its central role in electron transfer. Its diverse functional roles, along with the availability of high-resolution structural data since the early 1970s, have contributed to making this protein a paradigm in the study of electron-transfer processes (Pascher et al., 1996). It contains a c-type heme prosthetic group that is covalently bound to the polypeptide chain by means of two thioether bonds formed with the side chain of two cysteines, Cys14 and Cys17 (assuming horse heart numbering). The prosthetic group contains an iron atom that is coordinated by the four pyrrole nitrogens of the heme, by the Nε1 of Hys18 and by the Sγ of Met80 (Banci & Assfalg, 2001; Banci et al., 1999). The iron presents two physiologically relevant oxidation states: Fe^{2+} and Fe^{3+} (Banci & Assfalg, 2001).

Alignment analyses of mitochondrial cyt *c* sequences performed on different eukaryotes have shown that primary structure is well preserved during evolution; particularly, the most conserved region appears to be that around the axial Met80 ligand (Banci et al., 1999). The molecular native folding is such that only 7 to 10% (depending on the different cyt *c* considered) of the surface area of the prosthetic group is exposed at the protein surface (Brayer & Murphy, 1996).

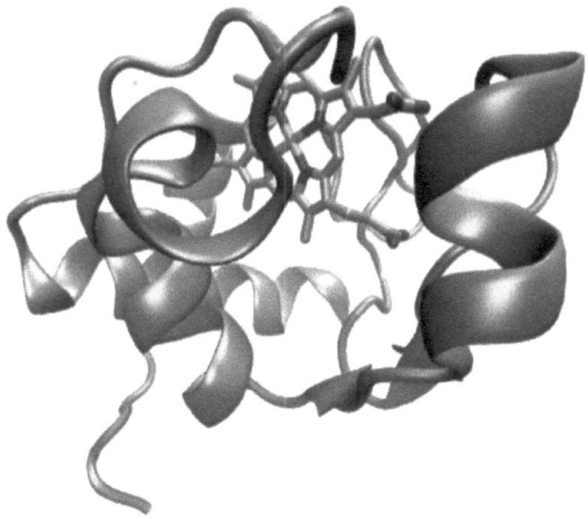

FIGURE 4.2

Schematic representation of the structure of cytochrome *c* from *Saccharomyces cerevisiae*. Note the heme structure in the active site of the molecule. Structural data from PDB file 1YCC.

Not surprisingly, most of the hydrophobic residues are buried. All the internal residues in van der Waals contact with the heme macrocycle are constituted by bulky hydrophobic groups that help stabilize the overall protein structure; Figure 4.2.

Further stabilization of the native structure is guaranteed by an extensive network of hydrogen bonds. As expected for a soluble protein, most of the charged groups are located at the protein surface, where their distribution appears highly asymmetric (Koppenol & Margoliash, 1982). Lysines are clustered mainly around the exposed heme edge and on the lateral sides of the molecular structure, whereas carboxylates are located primarily at the top and back. A comparative study of amino acid sequences including over 100 mitochondrial cyt *c* revealed that exposed residues are significantly more variable than buried ones. Therefore, a marked degree of conservation of specific exposed amino acids points to an important functional role in intermolecular recognition (needed for molecular docking with partners) and electron transfer.

Within the family of cytochromes *c*, iso-1-cyt *c* from *Saccharomyces cerevisiae* is the only member that contains a single partially surface-exposed cysteine (indeed it is located in a partially hydrophobic cage at the protein surface), which may be used for direct chemisorption on gold or other electronically soft metals (Gerunda et al., 2004).

4.2 Redox cofactors

A redox cofactor is a non-proteinaceous chemical compound that is bound to a metalloprotein and is required for the protein's functional activity. These proteins are often, but not exclusively (for instance, photosynthetic reaction centers), metalloenzymes, and cofactors can be considered as "support molecules" that assist the enzyme or the reaction centers in performing the biochemical transformations they are deputed to.

Cofactors can be either organic or inorganic molecules. They can also be classified according to how tightly they bind to an enzyme; loosely bound cofactors are named coenzymes, whereas tightly bound cofactors are termed prosthetic groups (e.g., the heme group). In the literature some authors prefer to use the term "cofactor" for inorganic substances, but we will not follow that route, since our focus is on organic, biology-related molecules. An enzyme depleted of its cofactors is called an apoenzyme, while the complete enzyme with cofactor is the holoenzyme (Sauke et al., 2001).

Some enzymes or enzyme complexes need several cofactors to function. For example, the multi-enzyme complex pyruvate dehydrogenase (Jordan & Patel, 2004), which exerts a key function at the junction of glycolysis and citric acid cycles, requires five organic cofactors and one metal ion: a loosely bound thiamine pyrophosphate (TPP), a covalently bound lipoamide and a flavin adenine dinucleotide (FAD), and the co-substrates nicotinamide adenine dinucleotide (NAD^+) and coenzyme A (CoA), and a metal ion (Mg^{2+}).

Organic cofactors are often vitamins or are made from vitamins. Many of them contain the nucleotide adenosine monophosphate (AMP) as part of their structures, such as ATP, coenzyme A, FAD and NAD^+. This common structure possibly reflects a common evolutionary origin as part of ribozymes in an ancient RNA-based world.

Cofactors can be classified into two broad groups: organic cofactors, such as flavin or heme (or chlorophyll and its bacterial counterpart bacteriochlorophyll), and inorganic cofactors, such as the metal ions Mg^{2+}, Cu^+, Mn^{2+}, or iron-sulfur clusters.

Organic cofactors are sometimes further sub-divided into coenzymes and prosthetic groups. The term coenzyme refers specifically to enzymes and, as such, to the functional properties of a protein. On the other hand, "prosthetic group" emphasizes the nature of the binding of a cofactor to a protein (tightly or covalently) and, thus, refers to a structural property. Some consider tightly bound organic molecules as prosthetic groups and not as coenzymes, while others define all non-protein organic molecules needed for enzyme activity as coenzymes, and classify those that are tightly bound as coenzyme prosthetic groups.

A 1979 letter in *Trends in Biochemical Sciences* noted the confusion in the literature and the essentially arbitrary distinction made between prosthetic groups and coenzymes and proposed the following scheme. Cofactors are defined as an additional substance apart from protein and substrate that is required for enzyme activity and a prosthetic group as a substance that undergoes its whole catalytic cycle attached to a single enzyme molecule (Bryce, 1979).

As we have mentioned, metal ions are common cofactors. The study of these cofactors falls in the general area of bioinorganic chemistry. In nutrition, the list of essential trace elements reflects their role as cofactors. In humans this list commonly includes iron, magnesium, manganese, cobalt, copper, zinc, selenium and molybdenum (Aggett, 1985). Although chromium deficiency causes impaired glucose tolerance, no human enzyme that uses this metal as a cofactor has been identified. Iodine is also an essential trace element, but it is used as part of the structure of thyroid hormones rather than as an enzyme cofactor (Cavalieri, 1997). Calcium is another special case, in that it is required as a component of the human diet, and it is needed for the full activity of many enzymes, such as nitric oxide synthase, protein phosphatases and adenylate kinase. It appears to activate these enzymes via an allosteric regulation, often binding to them in a complex with calmodulin (Clapham, 2007). Calcium is, therefore, a cell signaling molecule, and not usually considered a cofactor of the enzymes it regulates (Niki et al., 1996).

Other kinds of organism require additional metals as enzyme cofactors, such as vanadium in the nitrogenase of the nitrogen-fixing bacteria of the genus *Azotobacter* (Eady, 1988), tungsten in the aldehyde ferredoxin oxidoreductase of the thermophilic archaea *Pyrococcus furiosus* (Chan et al., 1995), and even cadmium in the carbonic anhydrase from the marine diatom *Thalassiosira weissflogii* (Lane & Morel, 2000).

In many cases, the cofactor includes both an inorganic and an organic component. One diverse set of examples is the heme proteins, which consist of a porphyrin ring coordinated to iron.

Iron-sulfur clusters are complexes of iron and sulfur atoms held within proteins by cysteine residues. They play both structural and functional roles, including electron transfer and redox sensing, as structural modules (Meyer, 2008).

Organic cofactors are small organic molecules (typically having a molecular mass <1000) that can be either loosely or tightly bound to an enzyme and directly participate in the reaction. In the latter case, when it is difficult to remove without denaturing the enzyme, it can be called a prosthetic group. It is worthy of note that there is no sharp division between loosely and tightly bound cofactors (Sauke et al., 2001). Indeed, many cofactors, such as, for instance, NAD^+, can be tightly bound in some enzymes, while being loosely bound in others (Sauke et al., 2001). Another example is thiamine pyrophosphate (TPP), which is tightly bound in transketolase or pyruvate decarboxylase, whereas it is less tightly bound in pyruvate dehydrogenase. Other coenzymes, such as flavin adenine dinucleotide (FAD), biotin and lipoamide, are covalently bound. Tightly bound cofactors are, in general, regenerated during the same reaction cycle, while loosely bound ones can be regenerated in a subsequent reaction catalyzed by a different enzyme. In the latter case, the cofactor can also be considered a substrate or co-substrate.

Vitamins can serve as precursors to many organic cofactors (e.g., vitamins B_1, B_2, B6, B_{12}, niacin, folic acid) or as coenzymes themselves (e.g., vitamin C). Many organic cofactors also contain a nucleotide, such as the electron carriers NAD and FAD, and coenzyme A, which carries acyl groups. Most of these cofactors are found in a huge variety of species, and some are universal to all forms of life. An exception to this wide distribution is a group of unique cofactors that evolved in methanogens, and are restricted to this group of archaea.

The class of cofactors which will be the focus of the next sections are those derived from quinones. Quinones are a class of organic compounds that are formally "derived from aromatic compounds (such as benzene or naphthalene) by conversion of an even number of –CH= groups into –C(=O)– groups with any necessary rearrangement of double bonds" (Nic et al., 2006). The class includes some heterocyclic compounds. Among quinones a paradigmatic one is 1,4-benzoquinone (Figure 4.3), which will be extensively discussed in electrochemical applications both in diffusion and as a surface-bound species (following suitable derivatization with various thiol-terminated chains; vide infra).

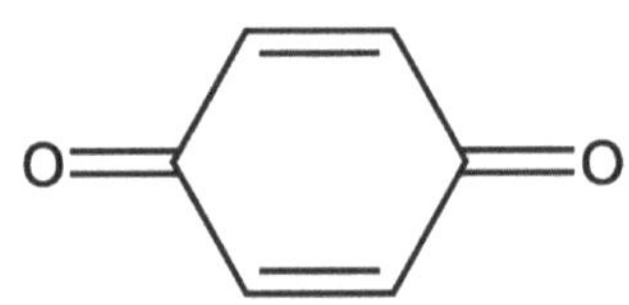

FIGURE 4.3

The structure of 1,4-benzoquinone.

Derivatives of quinones are common constituents of biologically relevant molecules. Some serve as electron acceptors in electron-transport chains such as those in photosynthesis (plastoquinone, phylloquinone) and respiration (ubiquinone) (Devault, 1984). Phylloquinone is also known as vitamin K_1 as it is used by animals to help form certain proteins, which are involved in blood coagulation, bone formation and other processes.

4.3 Driving redox reactions of freely diffusing molecules

Both redox metalloproteins and cofactors are naturally prone to be studied and affected by electrochemistry. We have seen how techniques like LSV and CV (Chapter 2) can be fruitfully used for retrieving information on both kinetics and thermodynamics of redox molecules in diffusion, and this is true also in the case of molecules of biological origin. Nonetheless, the application of direct electrochemistry to these classes of molecules requires some precautions and needs to overcome some difficulties connected to the effect that exposing, for example, a protein solution to a metal electrode can have on molecular integrity and functionality.

We are facing, in this case, two different kinds of problem. The first one, which essentially concerns proteins, is connected to the high free energy associated with the solid–liquid interfaces. These interfaces, usually, have the effect of inducing unwanted phenomena such as protein adsorption (with consequent possible electrode surface passivation) and partial or total protein denaturation, i.e., loss of protein native structure and consequent decrease or loss of functional activity. Both these eventualities have to be avoided if one is interested in either electrochemical characterization or electrochemical conditioning of a redox reaction.

The second one concerns the electronic coupling between redox center and electrode. It is, indeed, necessary that any molecule coming into contact with a working electrode from an electrolytic solution be in a condition to exchange electrons with it (both for measuring and affecting its redox state). Electron exchange requires that the molecule–electrode distance is short enough to allow for electron transfer and that the overlap between molecular orbitals and electrode ones is large enough to elicit that transfer. These two requirements are, somehow, coincident but it is worth quoting them both in view of the possible cases of large proteins or highly asymmetric cofactors that can physically come into contact with a working electrode, but not be prone to exchange electrons with it because of poor electronic coupling with the electrode. It can happen, for instance, that the redox site of a metalloprotein is buried inside a protein globule, or is located on the other side of the molecule with respect to that directly contacting the electrode. In both these cases, irrespective of the physical distance between the electrode and the molecule, no sizable electron transfer can take place.

In the next section we will see how all these problems have been circumvented in the case of redox metalloproteins, eliciting their direct electrochemical characterization. Then, we will review the main results shedding light on the electrochemical behavior of the benzoquinone/hydroquinone redox couple in solution. Those results will form the basis for the further development of our direct electrochemical approach to redox biomolecules, which will first require us to deal with redox metalloproteins and cofactors grafted onto a surface.

4.3.1 Redox metalloprotein electrochemistry in diffusion

Direct electrochemistry of redox metalloproteins is an intriguing approach for a number of reasons. First, the possibility of merging inorganic electrode materials with biological macromolecules bearing relevant catalytic and sensory properties appears appealing for potential applications. Second, as in the case of electrochemistry of simpler molecules, it appears possible to retrieve important information about intrinsic kinetic and thermodynamic properties even in the case of complex, biologically relevant macromolecules. Third, since, as we have mentioned in the previous section, electrical communication between proteins and electrodes can be established only once issues connected with protein denaturation, electrode passivation and proper surface docking have been settled, the successful achievement of protein direct electrochemistry will have far-reaching consequences.

Among the simplest redox metalloproteins, there are those whose role is just that of shuttling electrons between molecular partners. To this class belong relatively small (molecular mass in the range 5,000–20,000) proteins such as cytochromes, blue copper and iron-sulfur proteins. One could be tempted to think that due to their reduced size, the behavior of these proteins should bear a close resemblance to that of small redox molecules. In fact, the situation is quite different. Indeed, at variance with the outer-sphere electron transfer of small inorganic molecules, it has been shown by many studies that for small proteins the formation of a fast and reversible intimate, precursory, donor-acceptor complex docked so as to maximize electron transfer rate between partners is a requirement.

This requirement merges uncomfortably with the aforementioned tendency of proteins to undergo irreversible adsorption on metal electrodes coupled with conformational changes and loss of activity. This well-known phenomenon in bioelectrochemistry causes electrode contamination and passivation. The fact that nowadays it is almost routine to achieve reversible or quasi-reversible behavior in small protein voltammetry demonstrates that those problems have been solved.

The solution stemmed from efforts aimed at imparting to the electrodes' surfaces the same chemical properties relevant in macromolecule recognition, i.e., complementary hydrogen bonding, salt bridging and hydrophobic interactions (Armstrong et al., 1988). All these kinds of interactions, indeed, are important in a protein's contacts with its physiological partners.

The first clear demonstration of quasi-reversible direct electrochemistry of a metalloprotein appeared in 1977. Yeh and Kuwana (Yeh & Kuwana, 1977) studied horse heart cytochrome *c* and showed good CVs at indium-tin oxide (ITO) electrodes. Almost at the same time, Eddowes and Hill (Eddowes & Hill, 1977) demonstrated the quasi-reversible CV response of cyt *c* at a gold electrode, where a self-assembled monolayer of 4,4′-bipyridyl was formed.

The CV peak separation was about 60 mV and current peak intensities varied linearly with the square root of the scan rate. These figures indicate electrochemical reversibility and show that electron transport is dominated by diffusion to and from the electrode.

The two reported experiments have a common feature: the chemical groups terminating both electrodes' surfaces are able to interact specifically and reversibly with the protein surface.

As we have seen in section 4.1.2, the surface region surrounding the exposed heme edge, mechanistically relevant as the interaction domain, contains a number of charged primary amines (-NH_3^+) coming from lysine residues. Hydrogen bonding between these lysines and pyridyl nitrogens at the surface of the modified electrode appears to stabilize a transient electrode-protein complex, properly oriented to allow rapid electron transfer to and from the heme group (Eddowes & Hill, 1979a, 1982). Furthermore, electrochemical activity disappears as the pH is lowered from 7 to 4. Since the effective pK_a (5.1) of the pyridyl moieties is consistent with the protonation of one of the two nitrogens of 4,4′-bipyridyl, as a result it loses hydrogen-bond acceptor capability, replacing it with a (repelling) positive charge (Eddowes & Hill, 1979b). Furthermore, ellipsometric investigation of 4,4′-bipyridyl-coated gold shows that, whereas one nitrogen coordinates with gold, the other is left exposed, available for the interaction with the protein (Elliott et al., 1986).

This was just the first example of the use of so-called "promoters" of protein electrochemistry. Since then, extensive investigations have been carried out to find out the fine details defining a suitable promoter. Of course, this has to be chosen case by case, since its properties depend, in general, on both the electrode's nature and the protein properties.

Generally speaking, again in the case of cyt *c*, several different bifunctional molecules were used (Armstrong et al., 1988). As surface-adsorbing functional groups, the best performers were 4-pyridyl-N, -SH, S-S and P. All of them share the properties of being σ-donor, π-acceptor ligands, appropriate for coordinating to a group 11 metal like gold. In binding cyt *c*, the most effective functional groups could be divided into two categories: (i) neutral groups such as 4-pyridyl-N and aniline-type amines, which are weakly basic and able to form hydrogen bonds to lysine primary amines; (ii) anionic groups such as carboxylate, sulfonate, phosphonate, or phosphate that can interact with lysine-NH_3^+ through both hydrogen bonds and salt bridging. As to the group linking the two binding moieties, it can be flexible (e.g., aliphatic chain) or rigid (e.g., aromatic). Usually, rigidity shows superior performance but not at the level of the single molecule, rather, at that of the closely packed monolayer, so that closely packed aliphatic chains can perform as well as aromatic ones.

In the case of different proteins, as we mentioned earlier, one has to tailor the promoters in order to match protein-specific features. Unlike cyt *c*, for instance, most electron-transfer proteins show overall negative charges. Among these, there are bacterial flavodoxins and [4Fe-4S] ferredoxins, operating as electron carriers in nitrogen fixation, the chloroplast [2Fe-2S] ferredoxins, conveying electrons out of the photosystem I complex, and the blue copper protein plastocyanin, which is the essential electron donor in photosystem I in higher plants (Armstrong et al., 1988).

Hawkridge and co-workers (Landrum et al., 1977) succeeded in retrieving CVs from spinach chloroplast ferredoxin at a gold electrode modified with a polymeric coating of methylviologen.

Others used polyethylene glycol layers and the addition of $MgCl_2$ to promote stable electron transfer of different ferredoxins (e.g. from *Clostridium pasteurianum*).

The list of different redox proteins studied by the reported approaches continues to increase with time and it is beyond the scope of this book to provide an exhaustive account of applications in the field, which is periodically reviewed in the dedicated literature. What instead is important to us is to have established the feasibility of direct redox metalloprotein electrochemistry by the proper, rational chemical modification of the electrode surface. We will see, however, that another approach is available for obtaining a stable electrochemical response from redox proteins, and this is more prone to technological exploitation: direct chemisorption of electron-transfer metalloproteins onto the electrode surface. This approach will be described in section 4.4.

4.3.2 Redox cofactor electrochemistry in diffusion

In this section we will focus our attention on the peculiar electrochemical behavior of the benzoquinone/hydroquinone redox couple. This can be considered as a paradigmatic case of quinone-based cofactors, which, as discussed in section 4.2, play a key role in both photosynthesis and respiration. We are particularly interested in the electrochemical behavior of this redox couple, since it has a quite peculiar and complex redox chemistry, involving the exchange of two electrons and two protons. As such, on the one hand the interpretation of its electrochemical behavior is quite complex and has been elucidated by E. Laviron in a series of seminal papers (see, e.g., Laviron, 1984), on the other hand, its proton-dependent redox chemistry provides an intriguing opportunity to control kinetic and thermodynamic aspects of its reactivity by a bulk physical-chemical parameter, i.e., pH change.

Figure 4.4 shows a typical CV of the redox couple 1,4-benzoquinone/hydroquinone (1 mM in 50 mM H_2SO_4) at a polycrystalline gold electrode, from which one can readily see that the molecule behaves almost ideally in diffusion, featuring reversible oxidation and reduction waves, and an E_0' of 0.42 V (vs SCE).

The redox chemistry of this molecule has been shown to follow a quite complex reaction scheme. Particularly, the conversion between *p*-benzoquinone and hydroquinone can proceed along any path connecting the reacting intermediates of a nine-member square scheme (Jacq, 1971); Figure 4.5a. Each possible vertical step corresponds to a unitary change in molecular protonation, characterized by a given pK_a, whereas each horizontal one is related to a unitary exchange of electrons, with its rate constant and redox potential. In CV measurements these transformations, which can take place in principle though any of the pathways outlined in the square reaction scheme, are encompassed by just a pair of oxidation/reduction waves.

In a famous series of papers (see Laviron, 1984) E. Laviron showed that, under the assumption of equilibrium protonation (e.g., achievable in an aqueous buffer) and of a value for the symmetry factors of each electrochemical reaction equal to 0.5, the square scheme of Figure 4.5a reduces to a much simpler two-step one-electron reaction; Figure 4.5b.

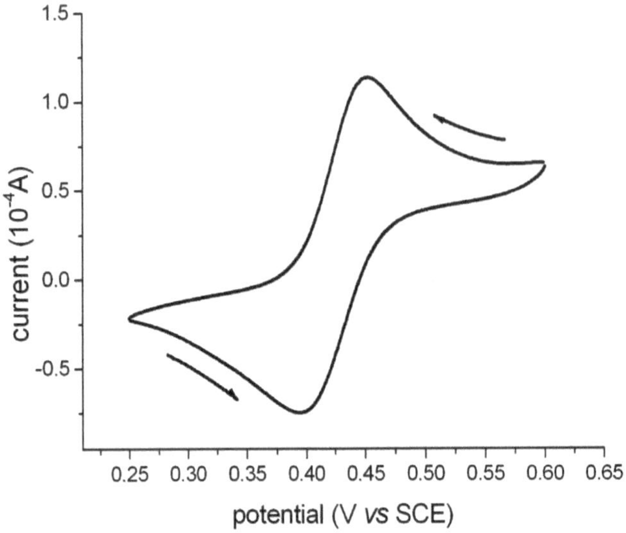

FIGURE 4.4

Cyclic voltammetry of the couple *p*-benzoquinone/hydroquinone at a polycrystalline gold electrode; concentration 1 mM in 50 mM H_2SO_4, scan rate 0.1 V/s.

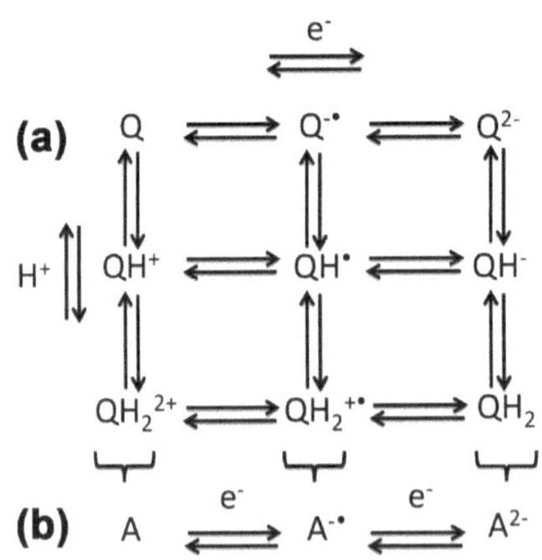

FIGURE 4.5

(a) Nine-member square scheme for the reaction between *p*-benzoquinone (Q) and hydroquinone (QH_2). (b) Equivalent reaction under equilibrium condition of protonation and symmetry factors of 0.5.

Furthermore Laviron, interpreting experimental data from Vetter (Vetter, 1952), showed that, under the aforementioned conditions, the trends with pH of the redox potentials for the two different reactions (Figure 4.5b), as extracted by Tafel plot measurements, are readily interpreted. They are constant in a pH interval spanning approximately 4–5.5. For higher (lower) pH values, whereas the redox potential corresponding to the first (second) transition of Figure 4.5b is constant, the other one decreases (increases), giving the corresponding plot a typical lozenge shape (Laviron, 1984). These results represent the theoretical and experimental reference point for any electrochemical characterization involving quinone or derivatized quinone molecules. They will be used for rationalizing the behavior of alkyl and oligo-phenylene-vinylene-derivatized *p*-benzoquinone, as we will see in section 4.4.2.

4.4 Driving redox reactions of surface immobilized molecules

There are several reasons to suggest that a redox molecule can be immobilized on the surface of a working electrode. As usual, these reasons can range from the requirements of fundamental investigation experiments to those of diverse applications. Among the former one can think, for instance, of studies aimed at understanding the role of the orientation and distance of molecules (proteins and prosthetic groups, cofactors) from the electrode's surface, whereas the second class encompasses all those applications where redox molecules are used together with metal or semiconductor electrodes to assemble hybrid structures for sensing, catalysis, transduction, etc.

Coating an electrode surface with a layer of redox molecules can even be regarded as an autonomous branch of science (surface bio-functionalization) and requires many specific attempts to succeed that can only partially be eventually exported to different molecular and electrode systems. Far from trying to be exhaustive on such a complex topic, here we just want to recall some principles that are common to different surface biofunctionalization approaches.

Assembling a (sub)monolayer of redox metalloproteins on an electrode's surface requires some preliminary knowledge of protein structure and function. The first aspect to consider is that proteins have to sustain the proximity of a metal surface without losing their structure and function. This, of course, depends upon fine details of the physical chemistry of both metal and protein surfaces and cannot be known in advance of some experiments. Generally speaking, however, the metal–liquid interface can induce conformational changes in proteins due to high interfacial surface tension, or even nonspecific protein adsorption, usually due to electrostatic interaction. Afterwards, one needs to choose a protein orientation. This again represents a major problem; whereas it is clear that the choice of protein orientation depends primarily upon the final goal for which the construct is assembled, it is generally complicated by the anisotropy of almost all proteins. In the case of a redox metalloprotein this anisotropy is mainly connected to the location of its redox active site inside the protein globule, the position and direction of the intramolecular electron transfer pathway(s), or the location of the docking regions for electron-transfer partners. In the case of a metalloenzyme, what matters is the location of the cage at the protein's surface that a substrate has to approach to be catalytically processed. Once this issue has also been settled, one comes to the choice of an immobilization strategy. This can be performed in a number of ways (Alessandrini & Facci, 2007), including the exploitation of exposed protein functional groups for direct binding to the electrode's surface, or to a bifunctional linker previously assembled as a monolayer on the surface, or by inserting some specific amino acids (usually cysteines) along the protein sequence by site-directed mutagenesis to yield a suitable anchoring point. This last aspect requires, of course, careful consideration of the expected effects on protein expression, purification, structure and function of the amino acid substitution/insertion at issue. Among commonly used linkers, if we consider a gold substrate, thiolated molecules are by far the most used. Their widespread use stems from the ease of direct chemisorption on gold, which, if coupled to a sufficiently intense inter-chain (attractive) interaction, can easily give rise to dense monomolecular layers at the surface. As to the second functional group of these bifunctional linkers, the choice depends upon the specific application, i.e., upon the target functional group on the protein.

Of course one has always to bear in mind that, since in direct electrochemical applications one wants to affect the redox state of an immobilized molecule by an electric potential imparted to a metal substrate (or, equivalently, measure a faradaic current arising from a redox reaction occurring at a substrate-immobilized protein), any intervening linker will result in the slowing down of the electron-transfer process between electrode and immobilized protein, tending to increase process irreversibility. From this consideration, the need arises to reduce to a minimum the linker length/resistivity by properly choosing the chemistry of the chain, which can be either unsaturated aliphatic (flexible but bad conductor), polyconjugated or aromatic (better conduction but rigid).

Other functional groups can be fruitfully used for anchoring molecules, hence proteins, at surfaces of both gold and electrodes of other materials. In section 4.3.1 we have seen the use of 4-pyridyl-N, and an S-S bridge in case of noble metal surfaces. Here it is just worth recalling the role of histidine tails, which, developed for binding Ni-NTA (nickel-nitrilotriacetic acid) grafted on surfaces, are also very useful for oxidized Ni and gold surfaces, as we will see in detail in Chapter 5.

The case of redox cofactors, such as the aforementioned paradigmatic case of benzoquinone, requires approaches similar to those described earlier for proteins. Particularly, it is either possible to make them react with proper functional groups (e.g. -SH, $-NH_2$, etc.) of bifunctional linkers suitable to chemisorption on the desired electrode by the other-end chemical moiety, or it is possible to synthesize them with a derivatizing chain of desired length, chemical composition and terminating functional group. We will see relevant examples of these approaches in section 4.4.2.

Once immobilization issues are settled, it is possible to characterize directly the protein film by direct electrochemistry. This is attained, for example, by measuring CVs of the immobilized layer. In this case the technique is called protein film voltammetry. At variance with diffusion protein voltammetry, protein film voltammetry is not affected by diffusion effects, which can complicate the interpretation of voltammetry signals. However, in real cases, the presence of linkers connecting proteins with the metal electrode, which unavoidably increase resistivity, and the particular protein orientation, which may not match the optimal one for fast electrode-protein electron transfer, usually give rise to quasi-reversible cyclic voltammograms.

Considering an ideal case, one has to deal with a reversible oxidation-reduction cycle upon voltage sweeping according to the general reaction:

$$protein^{ox} + ne^{-} \rightleftarrows protein^{red} \tag{4.1}$$

In this case, the reductive and oxidative waves are symmetrical, and the formal potential E_0' of the reaction (4.1) corresponds to the position of the maxima of the two waves. Figure 4.6 shows a scheme of such an ideal case.

Measuring the area under the oxidative and reductive waves allows one to know the number of exchanged electrons and, consequently, the number of electroactive molecules on the surface, provided that one knows the number of electrons involved in each elementary redox reaction. Indeed, after baseline subtraction (i.e., after getting rid of capacitance contributions to the

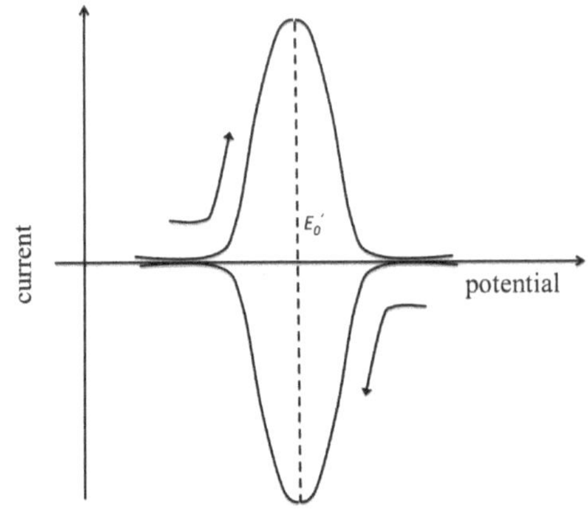

FIGURE 4.6

Ideal CV for an adsorbed redox species showing fully reversible behavior.

overall current), the faradaic contribution i_p (where $_p$ stands for "peak") is given by:

$$i_p = \frac{n^2F^2vA\Gamma}{RT} \tag{4.2}$$

where n is the number of exchanged electrons, F is the Faraday constant, v the scan rate, A the area of the electrode, Γ surface coverage.

As we have seen in section 2.6.2, the relation between the peak current and the voltage scan rate provides a useful clue on the adsorption state of the electroactive proteins. Adsorption is witnessed by a linear relation between these two parameters, at variance with the typical dependence upon the square root of the scan rate that is found in diffusion experiments (Bard & Faulkner, 2001). Ideally, for fully reversible electron transfer, the FWHM should be 3.53 RT/nF (i.e. 90.6 mV at 25 °C for $n = 1$) (Alessandrini & Facci, 2007). Nonetheless, deviations from this figure are possible due to the spread of redox potentials caused by nonhomogeneous distribution of the molecular adsorption modes on the electrode (Nahir & Boden, 1996). This phenomenon can take place even if proteins are adsorbed on a surface via a single linker that targets a unique protein functional group (e.g., the thiol moiety of a Cys residue) due to its possible flexibility coupled with a spread of protein orientation.

Similarly, protein film voltammetry can provide access to the standard electron transfer rate constant, k_S, of the protein-electrode reaction through analysis of the shift of the anodic and cathodic peak position as a function of voltage scan rate (Weber & Creager, 1994). In fact, if electron transfer is slow, a separation between oxidation and reduction peaks builds up for increasing scan rates (see section 2.6.2), which that can be used to extrapolate k_S, providing the electron-transfer rate of a zero free energy reaction (Weber & Creager, 1994).

This technique is particularly useful to determine the degree of electronic coupling between a redox metalloprotein and a metal electrode. In the next section, we will show how its use, along with that of site-directed mutagenesis for imparting different protein orientation on a metal surface via different locations of single reactive groups (e.g. cysteine) on the protein surface, allows one to establish the best electronic coupling among a set of different protein orientations.

Here, we conclude by noting that the aforementioned analysis can provide a detailed insight into electrode–protein coupling as mediated by distance when a redox protein is electrostatically adsorbed on a SAM of different chain length.

Cytochrome *c*, electrostatically adsorbed at the carboxyl acid headgroups of a SAM of alkanethiols on a gold electrode, shows a different electron-transfer behavior as a function of the number of methylene units n in the chain (Khoshtariya et al., 2003). For $n > 9$, i.e., at larger redox center–electrode distances, k_S shows an exponential dependence with the distance, with a decay constant increase of about 1 Å^{-1} per CH_2 unit. Decreasing the electrode–molecule distance by lowering the number of methylenes ($n < 9$), k_S features a plateau, suggesting the involvement of a different electron-transfer mechanism. These results have been interpreted either as a transition from a "non-adiabatic" regime of weak electronic coupling ($n > 9$) to an "adiabatic" one due to a stronger coupling of the redox center with the metal electrode ($n < 9$), or as due to a conformational modification of cyt *c* in closer proximity with the electrode, which limits electron transfer (Avila et al., 2000).

4.4.1 Effect of orientation at surface on electrode–protein coupling

Electrode–redox metalloprotein electronic coupling is a key parameter in designing efficient heterogeneous electron-transfer set-ups in the case of bioinorganic hybrid electrodes. We have seen in section 4.4 how this parameter can be quantified by analyzing the oxidation and reduction waves in CV measurements as a function of the voltage scan rate. In general, several factors can influence the electronic coupling when the electroactive molecule is a surface-immobilized protein. We have discussed previously the role of the distance from the electrode. One should also consider protein orientation and possible conformational changes induced by adsorption on a metal surface. Generally speaking, redox metalloproteins are anisotropic objects, and redox active sites are anisotropic too. These can be located inside a protein at different distances from the protein surface, being in general not in the protein center and having themselves an anisotropic shape, chemical composition, and ligation to the protein backbone. Therefore, it is extremely important to rely on tools that can vary protein orientation at the surface in order to evaluate the best coupling among those achievable. This tool, in the case of proteins, is protein engineering by site-directed mutagenesis. By mutating proteins it is possible to substitute selected amino acids at the protein surface with others, e.g., cysteine, that are very convenient for anchoring a molecule to a surface directly or through a reaction with a suitable linker molecule. Unfortunately, protein engineering is not a trivial or routine technique but suffers from important constraints that limit the possible amino acid substitutions in number, location along the polypeptide chain, and type (e.g., whether hydrophobic, polar or charged). For this reason, protein engineering is often assisted/guided by molecular dynamics simulations to predict plausible conformations of the mutated protein.

Now we will see how theoretical models have been developed that can help in predicting the electronic coupling also in the case of proteins adsorbed on a metal electrode in the framework of heterogeneous electron transfer.

According to theory (see section 3.7.1) the electron-transfer rate constant k between a donor and an acceptor can be expressed as the product of the square of the electronic coupling matrix H_{AD} (also called *transfer integral*) and a nuclear (or Frank-Condon) factor (FC):

$$k = \frac{2\pi}{\hbar}|H_{AD}|^2(FC) \tag{4.3}$$

Generally speaking, electronic coupling can depend on the properties of the intervening medium between donor and acceptor or, more precisely, on the different pathways that electrons follow to travel inside a molecule, from/to its redox site, to/from the electrode.

In order to compute the H_{AD}, fully *ab initio* methods are usually avoided due to the excessively large size of relevant proteins for these kinds of calculations. Recently, transfer integral calculations in the framework of density-functional theory (DFT) were proposed (Migliore et al., 2006).

Nevertheless, due to their simplicity, empirical models for the estimation of H_{AD} are widely used. Among those, a special mention is due to the tunneling pathways (Beratan et al., 1990, 1991, 1992) and average packing (Moser & Dutton, 1992; Page et al., 1999) density models, both of which provide predictions of relative or absolute electron-transfer rate constants.

In 1991, Beratan and co-workers proposed an empirical model known as "Pathways for estimating H_{AD}" (Beratan, Betts & Onuchic, 1991). This model treats the intervening medium between a donor and an acceptor as made of different segments each having its own tunneling decay factor β. In the case of intramolecular electron transfer, for example, the model factorizes the coupling in bonded, nonbonded and hydrogen-bonded steps that an electron should follow to go from a donor to an acceptor. Following this approach, the result is:

$$|H_{AD}|^2 = A^2\left(\prod_i \varepsilon_{bond}(i)\right)^2\left(\prod_j \varepsilon_{space}(j)\right)^2\left(\prod_k \varepsilon_{H-bond}(k)\right)^2 \quad (4.4)$$

where A is the value corresponding to the maximum coupling. It is easy to understand that the pathways model accounts for the large differences in through-bond and through-space electronic propagation. The unit-less decay factors corresponding to the different possible regimes are defined as:

$$\begin{aligned} \varepsilon_{bond} &= 0.6 \\ \varepsilon_{space} &= 0.5\varepsilon_{bond}e^{[-1.7(R_{DA}-1.4)]} \\ \varepsilon_{H-bond} &= \varepsilon_{bond}^2 e^{[-1.7(R_{DA}-2.8)]} \end{aligned} \quad (4.5)$$

with R_{DA} distance between donor and acceptor.

An alternative empirical model was developed by Dutton and co-workers, namely the average packing density tunneling model (Page et al., 1999). This model tries to balance through-bond and through-space decay of electron tunneling in proteins, writing for the electronic coupling matrix element as follows:

$$|H_{DA}|^2 = A^2 e^{\{-[0.9\rho+2.8(1-\rho)](R_{DA}-3.6)\}} \quad (4.6)$$

ρ being the packing density of protein atoms (varying in the range 1−0), defined as the fraction of the volume between donor and acceptor that lies within the van der Waals radius of the intervening atoms. The difference in the predictions of the average packing density and pathway models are generally not relevant when through-bond interactions dominate the electronic coupling.

Albeit developed for the interaction of freely diffusing proteins in solution, these methods can be fruitfully applied also to surface adsorbed molecules.

In a systematic study (Bortolotti et al., 2007), cyt *c* from *Saccharomyces cerevisiae* and horse heart, similar in structure and in redox properties, have been mutated in order to insert a single cysteine residue in different locations on the exposed surface of the protein. This approach enabled the possibility of direct chemisorption on gold electrodes and subsequent electrochemical characterization by CV measurements, aimed at measuring heme–substrate electronic coupling as a function of distance and prosthetic group orientation. In particular, the investigated molecules were the native recombinant and N62C in the case of yeast cyt *c*, and V11C, G45C and G56C for the horse heart molecule (Bortolotti et al., 2007).

Whereas the quasi-reversible CVs measured for the various mutants provided similar values of E_0', the resulting values for the apparent electron-transfer rate constant k_s were quite different; Table 4.1. Interestingly in the case of horse heart cyt *c*, k_s decreased with increasing distance between the heme-Fe and the C_α of each residue mutated into a cysteine. This behavior was in agreement with both theoretical predictions and experimental observations since the coupling matrix element is expected to decay exponentially with donor–acceptor distance.

However, the Euclidian distance between donor and acceptor is unlikely to be the only relevant parameter in determining the electron-transfer rate. Indeed, the N62C mutant of the yeast cyt *c* featured a larger k_s value than the native protein, despite a larger distance between heme and C_α of the

Table 4.1 Redox potentials for cyt *c* species covalently bound to an electrode through a (Cys)S-Au bond; calculated euclidean distances between the heme iron and C_α of the cysteine residue used for surface immobilization on gold; electrochemically determined rate constants for electron transfer between the heme and the electrode for the immobilized species. All the potentials are quoted vs SHE

Molecule	E_0' (mV)	Fe-CysC_α Distance (nm)	k_s^{-1} (s)
Horse heart V11C	+381	1.25	8.34
Horse heart G45C	+394	1.54	2.22
Horse heart G56C	+388	1.67	0.63
S. cerevisiae native	+370	1.39	0.20
S. cerevisiae N62C	+378	1.61	0.46

Adapted with permission from Bortolotti et al., J. Phys. Chem. C, *111, 12100–12105, 2007. Copyright 2007, American Chemical Society.*

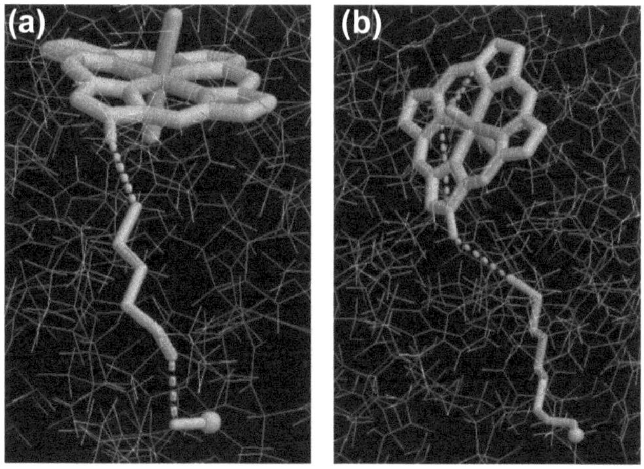

FIGURE 4.7

Best pathways for yeast cyt *c* and its mutant N62C, between heme cofactor and C_α of (a) yeast cyt *c*, Cys102; (b) N62C, Asn62.

Adapted with permission from Bortolotti et al., J. Phys. Chem. C, *111, 12100–12105, 2007. Copyright 2007, American Chemical Society.*

Cys residue. Therefore, other factors besides distance, such as intervening medium and protein orientation, appear to affect the rate of electron transfer.

The application of the tunneling pathway and average packing density model as implemented in the HARLEM software allowed an estimate of H_{AD} in a relative fashion. It turned out that the two different models for determining H_{AD} were both in acceptable qualitative agreement with experimental data and, as such, both suitable to predict the behavior of the electron-transfer rate constant of redox metalloproteins immobilized on a metal surface. Figure 4.7 shows, as an example, some of the most efficient pathways calculated from heme-Fe to C_α of Cys102 (for yeast cyt *c*) and Asn62 (for N62C mutant). Dotted segments correspond to through-space steps.

4.4.2 Redox cofactor electrochemistry at surfaces

A stable and robust surface linkage is a desirable and often necessary requirement also in the case of redox cofactor electrochemistry. Whereas, as in the case of redox metalloproteins, many of their fundamental properties can be investigated in diffusion experiments (see section 4.3.1), the reason for having them on a surface stems essentially from the idea of using their redox properties to implement electrochemically switchable hybrid surfaces. For such a goal, it is usually preferable to deal with a (sub)monolayer of molecules firmly anchored on an electrode surface rather than loosely assembled on it by milder means (e.g., physisorption, hydrophobic interaction, π-π stacking, etc.). That is why, in spite of the presence in the literature of works dealing with the electrochemical properties of surface layers of redox cofactors physisorbed on a surface, we will focus here on self-assembled (sub)monolayers of redox cofactors tethered to the surface by chemical linkers.

As already mentioned, we will focus on a special case of redox cofactors, derivatives of quinone, since they are paradigmatic cofactors that are particularly interesting because of the involvement of two electrons and two protons in the inter-conversion of the *p*-benzoquinone/hydroquinone redox couple (see section 4.3.2).

This fact makes them amenable to redox control on a metal surface coupled with the dependence of their electrochemical response on the solution pH.

One can follow different routes to obtain a (sub)monolayer of benzoquinone molecules grafted on a surface: exploit their inherent reactivity for suitable linkers, or derivatize them at synthetic level.

The first approach takes advantage of the reactivity of oxidized quinones towards nucleophiles (-SH, -NH_2, etc.). *p*-Benzoquinone can react twice with those groups, yielding a mono- and di-substituted molecules.

If that reaction involves functional groups of bifunctional linker molecules, it is a straightforward matter to assemble a (sub)monolayer of the reacted molecular complex. Indeed, such a layer can be assembled either by pre-forming a SAM of bifunctional linker molecules on an electrode surface (e.g., thiol functionalities on gold) or by reacting *p*-benzoquinone with bifunctional linkers in solution and, after reaction is completed, assembling the reaction product on a surface. The first approach is by far the one of choice, since it can easily take advantage of direct electrochemistry, imposing a suitable potential to a metal electrode pre-functionalized with a SAM of bifunctional linkers, to drive the redox state of surface-diffusing quinones to the oxidized form. Incidentally, we note that this approach is also suitable to enable space-resolved surface functionalization if one deals with independent metal pads as multiple working electrodes. This is often the case in various applications ranging from chemical and biosensors to molecular and biomolecular electronics (Alessandrini et al., 2008).

An alternative, more direct approach is that of synthesizing derivatized molecules with a proper tail terminating in a functional group that can be readily made to react with the metal surface of interest (e.g., −SH on gold).

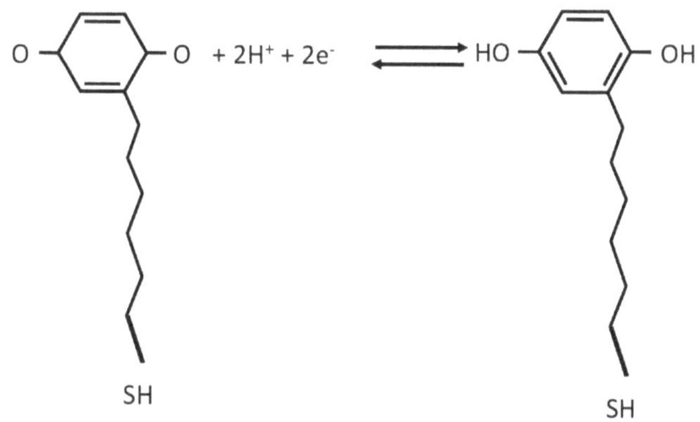

FIGURE 4.8

Structure and redox reaction of 2-(6-mercaptoalkyl)hydroquinone.

This case is exemplified by the synthesis of 2-(6-mercaptoalkyl)hydroquinone, whose structure and redox reaction is shown in Figure 4.8.

This molecule can be readily made to react with a gold electrode surface thanks to its terminal thiol moiety at the end of a 6-methylene chain and can therefore be arranged on a surface as a SAM.

The CVs of the SAM performed in 50 mM NH_4Ac, pH 4.6, display a typical quasi-reversible behavior, as revealed by peak separation (0.3 V at 0.1 V/s scan rate) and wave asymmetry. This is due to the presence of the alkyl chain connecting the benzene ring to the electrode, which slows down the heterogeneous electron-transfer process.

Applying Laviron analysis (see section 2.6.2) it is possible to verify that the measured electrochemical signal is that of an adsorbed monolayer, as the linear trend of the peak intensity with scan rate suggests. Furthermore, the trend of cathodic and anodic peak position as a function of ln(scan rate) provides clues about the number of electrons involved in the reaction. It turns out to be 1.6 rather than 2 as theoretically expected; the difference can be ascribed to a nonsymmetric energy barrier for the redox reactions involving the hydroquinone/quinone couple as already reported in the literature (Park & Hong, 2006; Park et al., 2009).

The proton-coupled electron-transfer reaction of the benzoquinone/hydroquinone redox couple shows electrochemical characteristics dependent on solution pH. This behavior bears the interesting consequence that electronic conductance of the molecules could be gated using pH changes (see section 4.5.4). Performing CVs at a fixed scan rate (0.03 V/s) as a function of pH in the range 4.6–7.6, it is possible to observe that the value of E_0'

scales with a negative slope with increasing pH, figuring an angular coefficient of (-54 ± 5) mV/pH, consistently with the expected value of -59 mV/pH for two electron/two proton reactions for a Nernstian redox species. At lower pH, the monolayer displays mainly hydroquinone moieties and the oxidation process involves an exchange of two protons and two electrons. As pH is raised, the monolayer goes partially into the deprotonated state and the reaction, correspondingly, involves a decreasing number of protons (one or none) and two electrons. Figure 4.9 reports the data described.

As we have seen, the electronic coupling, hence some aspects of the electrochemical behavior of these molecules, is determined by the alkyl linker featured by 2-(6-mercaptoalkyl)hydroquinone. It is instructive to observe what happens if one changes the electron-transport properties of the linker, for example by trying to increase the redox moiety electronic coupling with the substrate. A possible approach is to use a thiolated oligophenylene-vinylene chain to replace the alkyl one. This polyconjugated linker has much better electron-transfer properties with respect to the alkyl counterpart. Indeed, extensive studies have been reported in the literature (Salomon et al., 2003) presenting measurements of conductivity through di-thiolated organic chains (e.g., alkyl and OPV); they provide values for the corresponding β factor (the reciprocal of the tunneling decay length)

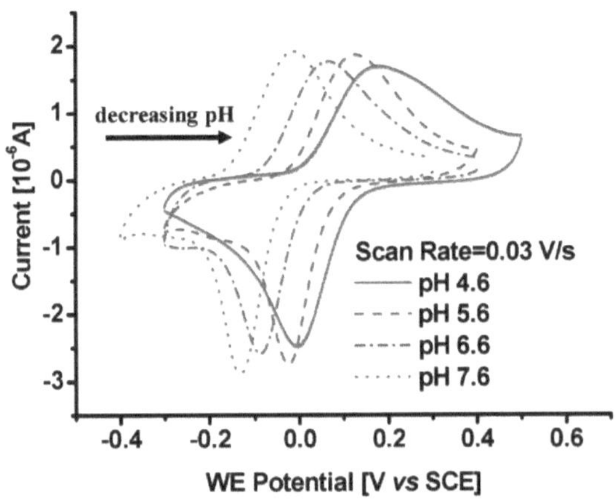

FIGURE 4.9

Cyclic voltammograms of a SAM of $H_2Q(CH_2)_6$-SH obtained at a scan rate of 0.03 V/s in 50 mM NH_4Ac for four different pH values: 4.6, 5.6, 6.6, and 7.6.

Adapted with permission from Petrangolini et al., J. Am. Chem. Soc., 132, 7445–7453, 2010. Copyright 2010, American Chemical Society.

of 0.6–1 $Å^{-1}$ for alkyl chains and 0.2–0.6 $Å^{-1}$ for OPV chains (Salomon et al., 2003), pointing to the role of polyconjugated chains in increasing molecular conductivity, hence electronic coupling with the metal substrate in the case considered here.

4-(2′,5′-dihydroxystyryl)benzyl thioacetate is a molecule that possesses the aforementioned characteristics. Figure 4.10 shows its structure. Apart from its OPV chain, the molecular structure is the same as 2-(6-mercaptoalkyl)hydroquinone. Particularly, its redox chemistry is the same as that of the alkylated counterpart.

CV measurements performed in the same conditions (pH 4.6, scan rate 0.5 V/s in 50 mM NH_4Ac) with respect to the alkylated molecules show the signature of an improved electronic coupling with the metal electrode. Indeed, peak separation decreases from ~400 mV in the case of the alkylated chain to ~100 mV in the case of the OPV chain. Figure 4.11 shows this comparison.

Furthermore, the dependence of peak position with respect to ln(scan rate) provides a figure of 3.4 s^{-1} for the apparent rate constant, to be compared with that for the alkylated molecule of 0.07 s^{-1}. This figure

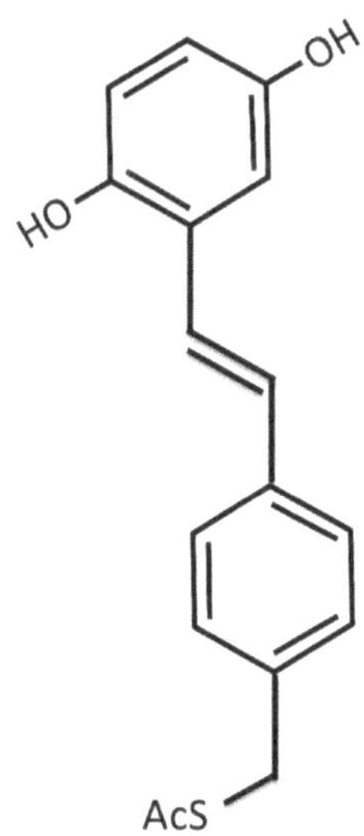

FIGURE 4.10

The structure of 4-(2′,5′-dihydroxystyryl)benzyl thioacetate.

Adapted with permission from Petrangolini et al., J. Phys. Chem. C., *115, 19971–19978, 2011. Copyright 2011, American Chemical Society.*

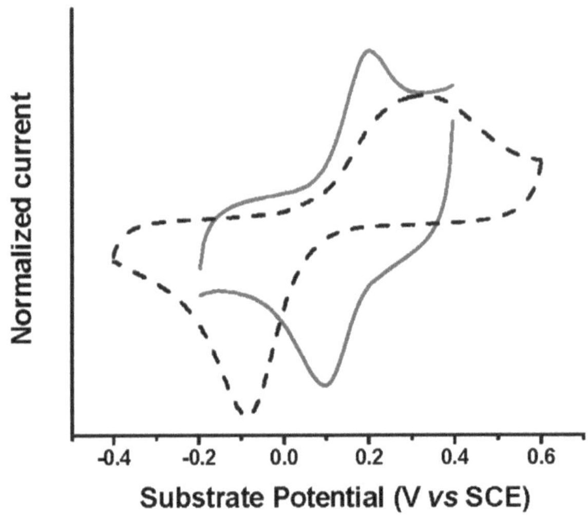

FIGURE 4.11

Comparison between CV measurements of alkyl (solid line) and OPV (dashed line) derivatized quinone monolayers at a gold substrate. Experimental conditions: 50 mM NH_4Ac, pH 4.6, scan rate 0.5 V/s.

Adapted with permission from Petrangolini et al., J. Phys. Chem. C., *115, 19971–19978, 2011. Copyright 2011, American Chemical Society.*

confirms, of course, the improved electronic coupling with the substrate. Also the dependence upon pH of the redox potential shows an angular coefficient of −59 mV/pH, in perfect agreement with both theoretical expectations and previous data on the alkylated counterpart.

The comparative electrochemical characterization of these two similar redox systems represents the starting point for the study of their behavior at the level of a single molecule, as we will see in section 4.5.4. That study will allow us to shed light on the electron-transport mechanisms underlying the heterogeneous exchange of electrons between these redox cofactors and metal electrodes.

4.5 Single biomolecule electron transfer

The interest in measuring and controlling redox reactions at the single biomolecule level stems on the one hand from the general motivations behind any single-molecule investigation and, on the other hand, from technological challenges and needs.

It is clear that bulk (electrochemical) measurements provide information on the average behavior of a large number of molecules. At that level one can get robust results describing the mean behavior of a molecular population that can, however, be quite heterogeneous. As such, the distribution of any relevant parameter will be affected both by homogeneous (i.e., dependent on the intrinsic properties) and inhomogeneous (i.e., dependent on the differences between molecular microenvironment, orientation, etc.) broadening. Hence, its mean value will reflect both these sources of uncertainty in a tangled fashion.

The single-molecule approaches allows one to get rid of inhomogeneous-broadening related effects, probing directly the "intrinsic" nature of the parameter at issue. It is furthermore evident that the distribution of events giving rise to a certain mean value contains much more information than the mean value itself. The information content of the single events giving rise to a certain distribution allows for a much deeper insight into the mechanisms underlying a certain phenomenon. In the case at issue, i.e., electron transfer in single biomolecules, having access to a large set of possible iso-energetic pathways that an electron can follow to reduce or oxidize a redox metalloprotein provides a unique clue to the possible physical-chemical mechanisms ruling the electron-transfer phenomenon.

There is, however, a further reason prompting single-protein study of electron transfer that is related to the possible technological exploitation of proteins. Indeed, proteins in general are the operative units of living beings, i.e., the molecules to which the most of the organisms' functional activity is deputed (Branden & Tooze, 1999). Despite the general statement that recognizes in the interplay between a protein and its environment the real responsibility for a given molecular conformation and function, one can affirm safely enough that proteins are the sites of functionality in living beings. As such, they represent the ultimate size units for a given biological function. Given their nanometer extent, it is straightforward to think of them in terms of artificial hybrid nanodevices whose functionality is intrinsic to them rather than emerging from the way the device is assembled (as in the case of more conventional, top-down devices such as electronic, solid-state ones). Therefore, studying and achieving control of biomolecule functionality at the single-molecule level emerges in the context of nanoscience as an extremely appealing possibility, that is, implementing single-molecule biodevices. These could not only enable an ultimate device size, but are also regarded as interesting for various features they can display. In the case of bio-electronics devices, i.e., devices that exploit the electron-transfer capabilities of a redox metalloprotein sandwiched between two nanometer-spaced electrodes, it is predicted that they should display typical power consumption as low as 10 pW, fast commutation times and large on/off ratios. Other charming possibilities are connected to the full exploitation of the self-assembly properties inherent to many biological systems, such as physiologically interacting partners, with the aim of achieving self-assembling (e.g., self-positioning) functional devices. It is straightforward to think that, for such single-molecule investigation and conditioning, one has to refer to tools coming from the realm of nanoscience and nanotechnology.

In the case most interesting for us, that of redox metalloproteins and cofactors, that tool is primarily the electrochemical scanning tunneling microscope (ECSTM).

4.5.1 Electrochemical scanning tunneling microscope

The advent of the scanning tunneling microscope (STM) (Binnig & Rohrer, 1983) can be considered as the starting act of nanoscience. The possibility of getting atomic resolution imaging in the direct space represented in the early 1980s an epochal breakthrough, destined to change suddenly and stably the way we studied surface phenomena while opening up novel, fascinating routes to the investigation of biological matter.

Soon after the initial excitement, it appeared clear that most of the biological samples were not amenable to being investigated by a technique requiring the establishment of an electron tunneling current flowing between a conductive substrate and a metallic probe, kept just a few angstroms apart. That is the why biological applications of scanning probe microscopes were almost completely monopolized by atomic force microscopy (AFM), which stemmed from STM technology in 1986 (Binnig et al., 1986).

The use of microscopic and spectroscopic techniques relying on electron tunneling measurements in the investigation of biological molecules was soon abandoned, mainly due to a deep lack of theoretical models for interpreting results, and remained almost irrelevant until the development of a novel approach, namely the electrochemical scanning tunneling microscope (ECSTM) (Siegenthaler, 1995).

ECSTM also makes use of a tunneling current to function, but differs from its progenitor STM in key aspects that make it a highly valuable tool for the investigation of redox molecules, including those of biological origin. Particularly, it can be operated in a liquid cell, under full potentiostatic control. These features not only enable one to perform STM imaging and tunneling spectroscopy in a liquid environment, but provide the technique with a further, unique control of the position of tip and substrate Fermi level that is unattainable by conventional, UHV operated, STM. The consequences of these special features, which will be soon analyzed in detail, have a major impact on the possibility of imaging and performing tunneling spectroscopy of biomolecules and redox molecules in particular. First of all, the presence of a liquid, water-based, salty environment allows one to probe biomolecular features in conditions that resemble closely those present in the physiological environment. This fact is, of course, a key aspect of the possibility of retrieving information on biomolecule conformation and function that can be readily related to their physiological state. Second, the possibility of performing STM/STS experiments under full potentiostatic control enables a unique investigation of the role of redox levels in enabling electron flow through a redox molecule.

Figure 4.12 shows a general scheme of an ECSTM. It outlines the main differences of this particular experimental set-up with respect to the normal UHV STM. Apart from the obvious observation that in this case no UHV apparatus is required and that the UHV environment is substituted by a liquid cell containing both substrate and tip, one can summarize the differences between ECSTM and STM in two main features that are present in the former and absent in the latter.

The first important feature that characterizes ECSTM is the use of a probe that is largely insulated. Particularly, the whole tip surface immersed in liquid has to be insulated, apart its very apex. This insulation is required in order to

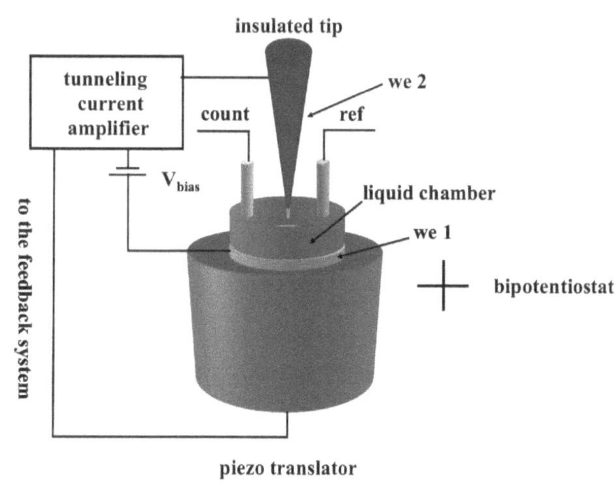

FIGURE 4.12

Scheme of an ECSTM. Note the four-electrode electrochemical cell, tip insulation, and the presence of a bipotentiostat.

prevent parasitic currents, capacitive in nature, entering the current amplification/feedback circuit while letting the tip apex sensing tunneling current. Indeed, for the technique to be operated as an STM, only tunneling current is to be used for controlling the tip-substrate position, thus contributing to the image formation mechanism or to any electron tunneling-related information retrieval. Tip insulation is a major issue in the ECSTM technique, not only for the role it plays in the technique, but also due to the difficulty in realizing it with a good success rate. There exist different approaches for insulating STM tips. The most common and widespread rely on coating a pre-sharpened tip with a layer of electrically insulating material such as a wax of a plastic polymer. It is often useful to assist such a procedure with an electrolytic deposition of an electropolymerizable adlayer, followed, for instance, by a melt wax film deposition. A common aspect to all the coating procedures is that the tip apex is de-wetted due to its extremely small radius of curvature. This feature, of course, is what makes these approaches so popular.

Nevertheless, the results obtainable by these methods are not strictly predictable and reproducible due to the inherent dependence of the final results on many variables that are usually poorly controlled (e.g., radius of curvature, temperature of melt wax film, speed of the coating process, environmental variables, etc.). Therefore, one needs to estimate the leakage current in the actual experimental conditions, that is, in a given electrolytic solution and at a given bias voltage, and at a tip—substrate distance large enough to prevent the tunneling current being sensed. Generally, depending on the application, in the bioelectrochemical context a leakage level that does not exceed 1—2 pA is considered acceptable.

The second specific feature of an ECSTM set-up is it relies on an external circuitry, namely a bipotentiostat, to drive the potential of the substrate and tip. In the measuring cell, indeed, four electrodes are placed: tip and substrate play the role of two working electrodes, then a reference and a counter electrode are needed. We have, therefore, a four-electrode electrochemical cell that requires a bipotentiostat to drive the potential of the two working electrodes independently.

The use of potentiostatic control of tip and substrate impinges on two different aspects of ECSTM operation: getting rid of faradaic currents and enabling tunneling spectroscopy in a liquid environment.

Faradaic contributions at either working electrode are in fact to be avoided if one is to measure tunneling currents flowing in the tip—substrate gap where a redox molecule can sit. The bipotentiostat enables the possibility of keeping the WEs' potential at values where only capacitive currents are present (hence, negligible because the tip is insulated) and no charge or mass transfer (e.g., adsorption of diffusing species) takes place. This is an important point since, if any faradaic contribution entered the current measuring circuit, the instrument could no longer operate as an STM.

It is common practice in ECSTM experiments to test the extent of the potential region where the tip shows just a capacitive behavior, by performing CV at the tip electrode while at a large distance from the substrate.

The second aspect is even more fundamental, since it involves a feature that is uniquely typical of ECSTM/STS and does not have any counterpart in conventional STM operation. As we mentioned above, a bipotentiostat can independently drive the potential of the two WEs (tip and substrate) involved in the four-electrode electrochemical cell; since a bias voltage is to be imposed between tip and substrate in STM operation, it represents the difference between tip and substrate potential (V_t and V_s, respectively):

$$V_{bias} = V_t - V_s \tag{4.7}$$

V_{bias} is usually kept constant during imaging in order to avoid charging/discharging the associated capacitor. Therefore, the bipotentiostat can rigidly shift V_t and V_s in the potential scale, that is, tip and substrate Fermi levels in the energy scale. This possibility is depicted in Figure 4.13 and is at the

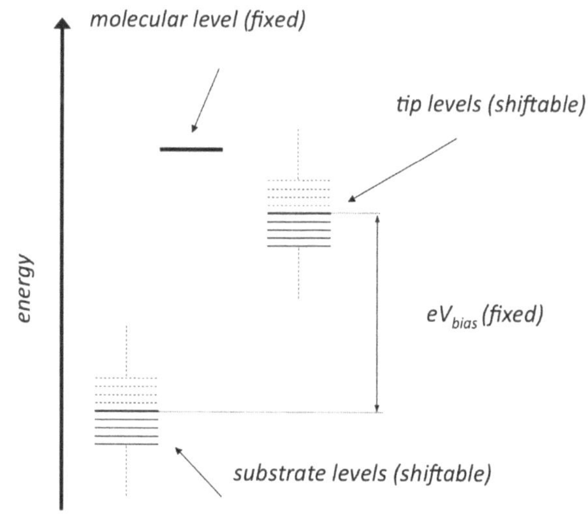

FIGURE 4.13

Energy diagram of an ECSTM. The bipotentiostat allows one to shift tip and substrate levels at constant V_{bias}, searching for molecular levels fixed in the energy scale.

basis of the so-called spectroscopy-like imaging mode of ECSTM. It consists in the possibility of imaging samples at different V_s (V_t) values, seeking for possible molecular levels, fixed in the energy scale, and potentially involved in the tunneling pathway between tip and substrate. Moving tip and substrate Fermi levels at constant V_{bias} enables, thus, a sort of spectroscopic imaging of molecular levels that is of fundamental relevance for assessing the role of the molecular (redox) level in the tunneling mechanism as mediated by the presence of a redox molecule.

We stress once more the difference between UHV STM, where only two electrodes are present (tip and substrate), and ECSTM. In the latter technique, the use of a bipotentiostat enables further control of the imaging/STS set-up than that provided simply by V_{bias}, making this technique a powerful tool for investigating the role of electronic levels brought about by redox (bio)molecules grafted on a surface and imaged in a physiological-like environment.

4.5.2 Theories for ECSTM of redox molecules

Along with the experimental development of ECSTM, various groups developed theories and models to interpret the conduction mechanism of redox adsorbates as probed by that technique. The proposed theories provide an expression of the ECSTM current as a function of the applied bias voltage, V_{bias}, and substrate potentials, V_s, that makes use of microscopic quantities connected to the molecule and to the molecule–electrode interaction.

Most of them are defined in the framework of the theory for (heterogeneous) electron-transfer reactions. If E_0 is the standard reduction potential for a redox couple, it is convenient to define the redox energy level of a molecule as:

$$\varepsilon_0 = -F \cdot E_0 \tag{4.8}$$

F being the Faraday constant. ε_0 represents, up to an unknown but fixed constant, the free energy of the process in which an electron is taken by an oxidized molecule in its equilibrium conformation and makes it a reduced one, again in the proper equilibrium conformation.

One finds it convenient to introduce a new quantity related to ε_0, which we call ε_0^{loc}. This is the redox energy level in the specific ECSTM experimental condition, i.e., it accounts for the fact that during ECSTM measurement the molecule is affected by the presence of tip and substrate (in particular it is affected by the electrostatic potential produced by those charged surfaces), by the electronic current, and by the confinement effects due to its being in a nanometer-sized gap. λ is, as usual, the reorganization energy for the molecule and represents the free energy that the molecule gains upon relaxation of the ET reaction products from the equilibrium conformation of the reagents.

In particular, we consider the reorganization energy of the ET reaction:

$$\text{Protein}^{n+} + \text{Electrode} \rightarrow \text{Protein}^{(n-1)+} + \text{Electrode} \tag{4.9}$$

where Electrode refers to either tip or substrate, according to the ECSTM/STS configuration adopted.

ε_{Fs} and ε_{Ft} are the electrochemical potentials of the substrate and the tip, respectively. They represent the boundary between occupied and empty electronic states in the metal, a boundary which is smeared at $T > 0$ K. To simplify the theoretical treatment, one can assume that the population of the electronic levels in the electrode is that for $T = 0$ K.

$T_t(\varepsilon)$ and $T_s(\varepsilon)$ are the electronic couplings (or transfer integrals) between the redox molecule and the electrodes (tip and substrate, respectively). Each level, within the continuum distribution in each electrode, has a different degree of coupling with the redox molecule, thus a specific value of $T_{t/s}(\varepsilon)$.

$\rho_t(\varepsilon)$ and $\rho_s(\varepsilon)$ represent the density of electronic states at energy ε in tip and substrate, respectively.

Now we move on, facing the description of the most representative among the proposed models for ECSTM conductivity and the corresponding equations $I_t(V_s, V_{bias})$. We will limit our treatment to those theories assuming a classical behavior of the nuclear degrees of freedom of the system, even if some works have evaluated the effects of a quantum behavior of such degrees of freedom (Schmickler, 1993).

This choice allows us to focus on the transport mechanism as described by various models.

In a chronological fashion, we start with the resonant tunneling model.

In 1992 Schmickler and Widrig proposed a model to explain the conductivity of redox adsorbates in the STM configuration (initially, disregarding potentiostatic control of substrate and tip) (Schmickler & Widrig, 1992). They assumed that the molecule redox level fluctuates due to coupling to a harmonic oscillator bath, treated classically (Schmickler, 1993), and they included the effect of such a bath in terms of reorganization energy and (broadened) density of redox states.

In their model, electrons tunnel from tip to substrate (or vice versa, depending on the sign of V_{bias}) through the empty redox level of the molecule. Tunneling is only effective when the initial, occupied tip electron level, the intermediate molecular level and the final, empty substrate level all have the same energy (resonant tunneling). Under the further assumption that the molecule–electrode electronic coupling is small with respect to the reorganization energy, the expression for the electron current results (Schmickler, 1993; Schmickler & Widrig, 1992; Schmickler & Tao, 1997):

$$i = -\frac{F\pi}{\hbar}\int_{\varepsilon_{Fs}}^{\varepsilon_{Fs}-FV_b}\frac{\rho_t(\varepsilon)|T_t(\varepsilon)|^2\rho_s(\varepsilon)|T_s(\varepsilon)|^2}{\pi\left[\rho_t(\varepsilon)|T_t(\varepsilon)|^2+\rho_s(\varepsilon)|T_s(\varepsilon)|^2\right]}D_{ox}(\varepsilon)d\varepsilon$$

$$D_{ox}(\varepsilon) = \sqrt{\frac{\pi}{k_BT\lambda}}\,e^{-\frac{\left(\varepsilon-\varepsilon_0^{loc}-\lambda\right)^2}{4k_BT\lambda}} \qquad (4.10)$$

where $D_{ox}(\varepsilon)$ is the density of oxidized states, after Gerischer (Gerischer, 1960), which accounts for the probability distribution of the fluctuating redox level.

When the molecular redox level is occupied (reduced molecule), the theory can be reformulated in terms of hole transfer. The resulting expression is analogous to (4.10), where $D_{ox}(\varepsilon)$ is replaced by $D_{red}(\varepsilon)$, the density of reduced states.

It is noteworthy that, even if the physical mechanism at the basis of (4.10) is resonant tunneling, i.e., electrons never localize on the protein redox level, λ appears in the expression for the current. This happens for two reasons: (i) λ is a measure of the thermal fluctuations of the redox level energy; (ii) λ can also be interpreted as the difference between the energy of the molecular level available for the incoming charge (i.e., the level that matters in resonant tunneling) and ε_0^{loc}. Thus, $D_{ox}(\varepsilon)$ displays a maximum for $\varepsilon = \varepsilon_0^{loc} + \lambda$, and a width around this maximum which is connected to λ. The behavior of the current as a function of V_s at fixed bias voltage can be easily extracted using (4.10). In particular, making the assumption that $T_t(\varepsilon)$ and $T_s(\varepsilon)$ change negligibly in the region around the explored ε_{Fs}, $I_t(V_s)$ shows a maximum for $V_s = V_s^{max}$. It is given by:

$$V_s^{\max} = -\left(\varepsilon_0^{loc} \pm \lambda\right)/F - V_{bias}/2 \qquad (4.11)$$

i.e., when the maximum of $D_{ox}(\varepsilon)$ (+ sign) or $D_{red}(\varepsilon)$ (– sign) is halfway between tip and substrate potential.

In the original Schmickler treatment, an implicit assumption was made that electrons never get trapped on the molecule, even if, during the measurement time, a sizable number of them cross the molecule. Actually, there is a finite probability for the instantaneous redox level to have an energy below both ε_{Fs} and ε_{Ft}, i.e., to be in a region where tunneling is prohibited (no empty levels available) and both electrodes are prone to reduce (i.e., give electrons to) the molecule. Since adsorbate fluctuations are slow with respect to electron tunneling, but not with respect to the macroscopic measurement duration, such regions are possibly explored by the system during the measurement. This point is indeed taken into account in the more recent stochastic adiabatic model by Kuznetsov and Schmickler (Kuznetsov & Schmickler, 2002), where the mean population of the redox level (varying between 0 and 1) adapts to the instantaneous value of the redox-level energy. Interestingly, in this recent model V_s^{max} appears to be close to $-\varepsilon_0^{loc}/F$ for small V_{bias}.

Resonant tunneling requires that electron transfer between tip and substrate is much faster than the typical fluctuations of the redox-level energy. One can consider, in contrast, the situation where electron transfer is much slower than typical redox energy fluctuations (typically, when the electronic couplings of the redox center with tip and substrate are small). In this limit, the overall tip-to-substrate electron transport can be regarded as the result of two processes that occur sequentially: firstly, one ET from the tip to the redox molecule, secondly, another ET from the molecule to the substrate. The second process does not preserve memory of where the molecular electrons come from, i.e., there is no correlation between the two processes. The calculation of the overall current reduces to a standard chemical kinetics problem, involving electrochemical rate constants (Zhang et al., 2002):

$$i = F\frac{k_{rt}k_{os} - k_{rs}k_{ot}}{k_{rt} + k_{rs} + k_{ot} + k_{os}} \tag{4.12}$$

where k_s represents the rate constants of the reduction (r) or oxidation (o) of the redox molecule by the tip (t) or the substrate (s). The dependence of I_t on V_s and V_{bias} comes from that of $K_{r/o,s/t}$ on such potentials. When the electronic couplings between tip and substrate are of similar magnitude, it can be estimated that:

$$V_s^{max} = -\varepsilon_0^{loc}/F - V_{bias}/2 \tag{4.13}$$

This relation indicates that maximum is achieved when the redox level of the molecule is midway between the Fermi levels of the two electrodes. As noted by Kuznetsov and coworkers, (Zhang et al., 2002) eq. (4.12) for the current implies that the ET reaction between the two electrodes and the redox molecule is non-adiabatic. Non adiabaticity, in this context, requires that (Zhang et al., 2002):

$$\mathrm{K_{el}}\rho_s(\varepsilon_s)\mathrm{KT} \ll 1 \quad \mathrm{K_{el}}\rho_t(\varepsilon_t)\mathrm{KT} \ll 1 \tag{4.14}$$

where K_{el} is the electron transmission factor. If a system is in the opposite full-adiabatic regime, Kuznetsov and coworkers introduced a different mechanism for electron transport. They considered an initially oxidized molecule that, due to thermal fluctuations, receives an electron from the more negative electrode. The molecule starts relaxing toward its reduced equilibrium configuration, but, since ET is much faster than relaxation, during this process several electrons can flow between the two electrodes, effectively amplifying the current. Kuznetsov and coworkers proposed that in this fully adiabatic limit, for $|FV_{bias}| < \lambda$ and $T_s(\varepsilon_s)\ \rho_s(\varepsilon_s) \sim T_t(\varepsilon_t)\ \rho_t(\varepsilon_t) = T\rho$, the current is given by (Zhang et al., 2005; Albrecht et al., 2005):

$$i = \frac{1}{2} F(FV_{bias}) K_{el} \rho \frac{\omega_{eff}}{2\pi} \exp\left(-\frac{\lambda - FV_{bias}}{4k_B T} \right) \cosh^{-1}\left[\frac{\frac{1}{2} FV_{bias} + \varepsilon_0^{loc} - \varepsilon_{Fs}}{2k_B T} \right] \tag{4.15}$$

where ω_{eff} is the effective vibrational frequency of all the nuclear modes that contribute to λ. Again, when the electronic couplings between the molecule and the two electrodes are similar, V_s^{max} from (4.15) fulfils (4.13).

Finally, for the non-adiabatic case, another model has been proposed by Kuznetsov and coworkers (Friis et al., 1998) which is in some sense intermediate between the resonance tunneling and the sequential two-step model. In fact after the first ET from the more negative electrode to the molecule, this starts to relax, as in the sequential two-step model. However, differently from this model, the second ET (from the molecule to the positive electrode) takes place before the complete relaxation of the system. Assuming that the vibrational coherence is not lost between the first and the second ET events, the expression for the current is (Friis et al., 1998):

$$i = \frac{\pi F}{\lambda \hbar^2 \sqrt{kT |U_s^* - U_t^*|}} \int_{\varepsilon Fs}^{\varepsilon Ft} d\varepsilon_t \rho(\varepsilon_t) |T_t(\varepsilon_t)|^2 \exp\left(-\frac{\left(\lambda + \varepsilon_0^{loc} - \varepsilon_t\right)^2}{4\lambda kT} \right) \times \int_{\varepsilon Fs}^{\varepsilon t} d\varepsilon_s \rho(\varepsilon_s) |T_s(\varepsilon_s)|^2 \tag{4.16}$$

where we have assumed that the tip is negatively-biased (in the opposite case, just exchange t with s). U_s^* and U_t^* are the free energy of the lowest transition states for the first and the second ET. As usual, when the initial condition of the molecule is the reduced state, everything can be re-formulated in terms of hole transfer. One can note (Friis et al., 1998) that (disregarding the pre-integral factor):

$$V_s^{\max} \approx -\left(\varepsilon_0^{loc} \pm \lambda\right)/F \qquad (4.17)$$

where the symbol $\pm$ refers to electron or hole transport. Similarly to what we noted in the case of resonant tunneling, for the vibrational coherent model it is also implicitly assumed that electrons (or holes) never get trapped in the molecular redox level.

Those reported are in synthesis the available theories, to date, for interpreting ECSTM data from both redox metalloproteins and redox cofactors and they will be used in the remainder of this chapter for extracting quantitative information out of data on both classes of samples. It is here just worth stressing the fact that an assessment of the proper electron-transfer mechanism is usually not simple in the case of spectroscopy-like imaging data and a safer attempt can only be done in the case of tunneling spectroscopy evidence.

4.5.3 Application to redox proteins

An application of ECSTM spectroscopic imaging that is paradigmatic both of the potential of this microscopy technique and of the behavior of redox metalloproteins in a tunneling gap is the study of the blue copper protein azurin.

As we have seen in section 4.1.1, azurin chemisorbs readily on gold via its disulfide bridge, which, in proximity to the metal surface, reduces and forms two S-Au bonds.

Atomically flat (e.g., Au(111)), azurin-coated gold can be imaged by ECSTM, taking advantage of the possibility of performing spectroscopic imaging as a function of substrate potential. This kind of experiment aims at investigating the potential role of redox levels associated with the molecule, i.e., with the copper ion in the protein's active site, in participating in the electronic conduction via the molecule itself.

The possible scenarios emerging from these experiment are different, involving the relation between several relevant parameters, vide infra, and can be rationalized on the basis of theories present in the literature (section 4.5.2) (Alessandrini et al., 2006).

A first example, which represents also the first published evidence of substrate potential dependent STM contrast for a metalloprotein (Facci et al., 2001), reports the variation with substrate potential of the apparent height of azurin measured with respect to a non-varying substrate reference level.

Imaging a sample region as a function of V_S in the potential range (−300−+100) mV (vs SCE), surface corrugation appears to change quite abruptly for values > −100 mV, revealing the presence of surface bumps; Figure 4.14. One can observe that for large enough tunneling current set points (e.g., 2 nA at V_b = +400 mV), azurin's apparent height follows a sort of sigmoidal trend with V_S. It is worth noting that the apparent bumps are electronic in nature, since if one performs the same experiment with a different kind of scanning probe microscope such as an AFM, non-sensitive to tunneling current but providing sample topography, the corresponding bumps are present in the whole potential range and do not depend upon V_S.

The picture apparently becomes more complicated if one performs a slightly different experiment where the current set-point is lowered (1 nA) and the apparent height variation is now measured against an internal standard, i.e., Zn-azurin, where native copper has been replaced by zinc, which it is well known not to change its oxidation level in the potential windows used in the

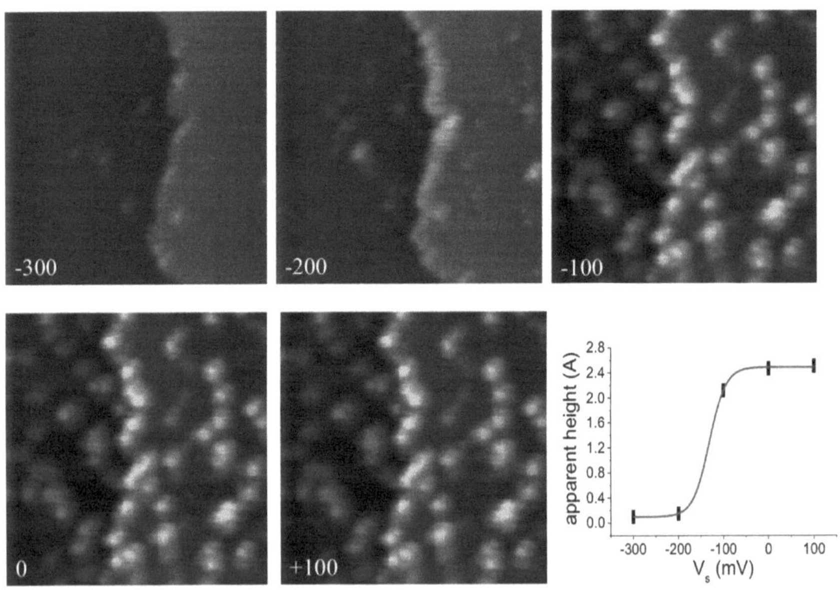

FIGURE 4.14

Set of ECSTM images of azurin on Au(111) acquired as a function of V_S (indicated on each image vs SCE). Imaging conditions were: I_t = 2 nA, V_{bias} = +400 mV, imaging buffer 50 mM NH_4Ac, pH 4.6. Note the sigmoidal shape of the trend of the apparent height as a function of V_S.

experiment. Zn, particularly, cannot undergo a reaction $Zn^{2+} + e^- \rightarrow Zn^{1+}$ but only $Zn^{2+} + 2e^- \rightarrow Zn^0$ at a much lower potential (~ -0.7 V vs SHE). Therefore, as has been directly demonstrated (Alessandrini et al., 2003), Zn-azurin does not change its apparent height with V_s in the range $-0.5 - +0.1$ mV.

In this case, the apparent height variation of the molecules changing their height with respect to the inert one shows a resonance-like behavior, as reported in Figure 4.15.

Azurin appears to behave in a different way in the two experiments, featuring a sigmoidal or a resonance-like trend according to the modification of the current set point (from 1 nA to 2 nA).

This apparent discrepancy can be rationalized in the framework of the theories describing the specific case of a redox protein in an ECSTM gap. Particularly, the two-step coherent electron transfer model by Kutznezov and Ulstrup (Friis et al., 1998) predicts qualitatively the exact trends observed in the two previous experiments. Their difference is traced back to that in magnitude between eV_{bias} and λ. If $eV_{bias} > \lambda$ theory predicts a plateau in the current vs V_s plot, whereas it predicts a resonance-like curve when $eV_{bias} < \lambda$; Figure 4.16.

The reason why the two experiments agree with these different trends can be ascribed to the variation of λ brought about by a different current set-point. Indeed, different set-points in STM mean a different tip–sample distance. Such a difference is generally marked for 1 nA variation (although difficult to estimate quantitatively due to the lack of information on the value of β at the protein's location) due to the exponential dependence of the tunneling current upon the tip–sample separation. One can argue that the solvent-related component of λ can be affected in a major fashion by changes in the vertical position of the tip (a macroscopic object on the molecular scale). These movements result in a sizable change in the value of λ that, in its turn, can toggle the system between the two different regimes.

It is fair to note, however, that such an explanation, although plausible, operates just at a qualitative level. This is because any theory of ECSTM phenomena provides predictions on $I_t = I_t\ (V_s, V_{bias})$, and not on the apparent height variation. In order to be more quantitative, one has to get direct access to the tunneling current; this, however, requires a different experimental approach, which will be introduced in section 4.6.

It is important to point out the fact that ECSTM experiments allow one to distinguish between redox metalloproteins having the same conformation and differing just in the electronic levels brought about by the (buried) metal ion. Figure 4.17 shows the remarkable difference between brighter bumps,

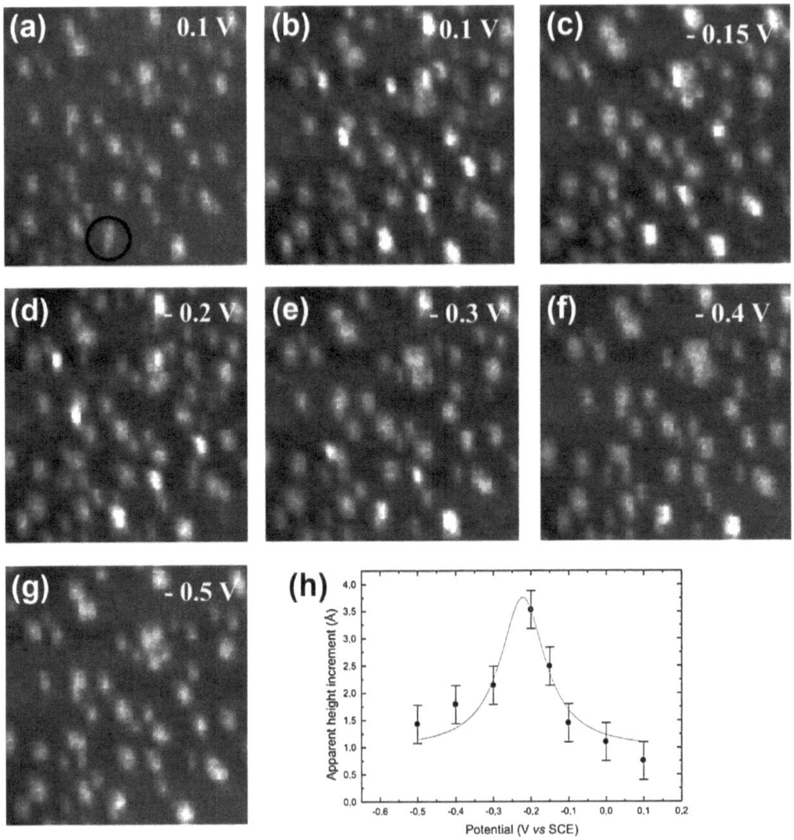

FIGURE 4.15

Set of ECSTM images (a-g) of azurin on Au(111) acquired as a function of V_S. Imaging conditions were: $I_t = 1$ nA, $V_{bias} = +400$ mV, imaging buffer 50 mM NH_4Ac, pH 4.6. Note the resonance-like shape of the trend of the apparent height as a function of V_S, measured vs those bumps that do not vary their apparent height (h).

From Alessandrini et al., Phys. Chem. Chem. Phys., *8, 4383–4397, 2006. Reproduced by permission of the PCCP Owner Societies.*

ascribable to Cu-azurin, and darker ones, corresponding to the non-active Zn counterpart.

4.5.4 ECSTM of benzoquinone/hydroquinone couple

ECSTM has shown its potential also in the single-molecule investigation of redox cofactors such as quinone-based molecules. We have discussed in

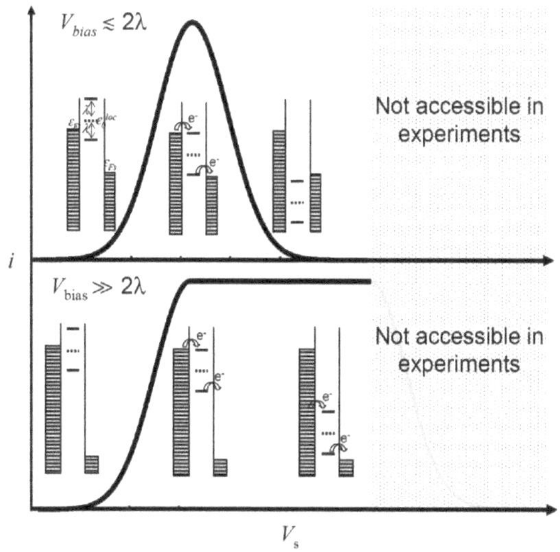

FIGURE 4.16

The two different trends predicted by two-step electron transfer theory.

From Alessandrini et al., Phys. Chem. Chem. Phys., *8, 4383–4397, 2006, RCS Publishing. Reproduced by permission of the PCCP Owner Societies.*

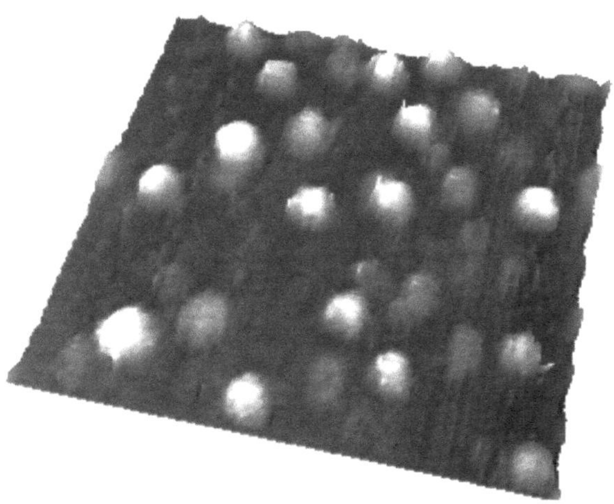

FIGURE 4.17

ECSTM image of a mixture of Cu (brighter bumps) and Zn-azurin (darker bumps). Imaging conditions were: $I_t = 1$ nA, $V_{bias} = +400$ mV, $V_s = +200$ mV vs SCE, imaging buffer 50 mM NH_4Ac, pH 4.6. Image size 89×89 nm^2.

sections 4.2, 4.3.2 and 4.4.2 the quite complex, pH-dependent, redox chemistry that characterizes these molecules. They are representative of cofactors that play a prominent role in biological ET, but they are also very intriguing for their potential application as multi-level molecular switches in view of molecular electronics implementations.

In the case of these molecules, ECSTM shows a substrate potential dependent STM apparent height (contrast) variation that is disclosed by the appearance of two regions of contrast enhancement (Petrangolini et al., 2010), at variance with the case of azurin, discussed in the previous section, where only one of these regions was evident. This behavior can be qualitatively observed both in the case of 2-(6-mercaptoalkyl)hydroquinone and in that of 4-(2′,5′-dihydroxystyryl)benzyl thioacetate. Its cause is traced back to the presence of two redox levels that can separately contribute to the tunneling current via the molecule. Whereas the analysis of ECSTM images indicates just a qualitative agreement with the theory, as we have seen in the case of azurin, a much more informative experiment is that which provides direct access to the tunneling current, enabling, thus, quantitative comparison with theories. The approach followed in the case of quinone derivatives has been that of assembling dense SAMs and then measuring tunneling current as a function of substrate potential V_S. This type of measurement requires the STM feedback to be disabled (or kept at a minimum, just for compensating for possible drifts), after having locked the desired current set-point. Figure 4.18 reports the corresponding current profiles in the case of quinone derivatives with the two different linkers. Both of them show two resonances; interestingly, those measured on OPV-derivatized quinone are much more closely spaced in energy (voltage), pointing to a better coupling between the redox moiety and the metal surface.

In both cases, similarly, a pH change causes a modulation of the distance between the two regions of tunneling current enhancement. Particularly, by increasing pH from 4.6 to 7.6, the peak distance decreases markedly (from 400 mV to 100 mV in the case of OPV-derivatized quinone). This behavior is readily interpreted in the framework of Laviron theory (see section 2.6.2) and supports the idea of a pH control on the functional behavior of these molecules. These findings point to the stabilization of the redox intermediate that is not visible in the case of CV measurements, while they reveal a separate ET behavior for the two redox reactions to which the nine-membered square matrix of section 4.3.2 condenses in the case of equilibrium protonation. Increasing pH decreases H^+ availability to the reaction, and can be regarded as another way, further than substrate potential, to modulate the quinone redox state.

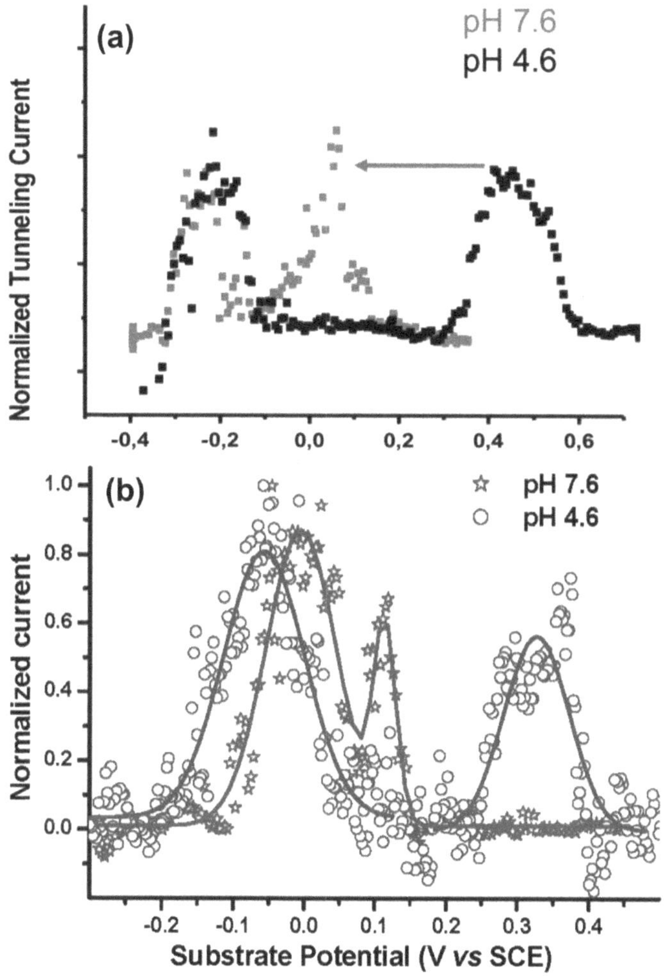

FIGURE 4.18

Tunneling current as a function of V_S in the case of benzoquinone derivatized with (a) alkyl chain, (b) OPV chain. Both measurements at pH 4.6 and 7.6 are reported for the two different linkers. $V_{bias} = +100$ mV, $I_t = 0.5$ nA.

Adapted with permission from (a) Petrangolini et al., J. Am. Chem. Soc., *132, 7445–7453, 2010. Copyright 2010, American Chemical Society; and (b) Petrangolini et al.,* J. Phys. Chem. C., *115, 19971–19978, 2011. Copyright 2011, American Chemical Society.*

Besides, it is worth noting that the close proximity of redox levels in the case of OPV-derivatized quinone at higher pH enables us to test theoretical predictions about the behavior of two-level redox molecules studied by

ECSTM. Indeed, in a work by Kuznetsov and Ulstrup (Friis et al., 1998) a different trend of tunneling current as a function of V_s is predicted if two redox levels are spaced by less than eV_{bias}. In that case, the double peaks of Figure 4.18 modify to a double plateau trend as a function of V_s due to the contemporary involvement in the energy windows defined by eV_{bias} (Petrangolini et al., 2011).

4.6 Electrochemically gated single-protein transistor

We are now in a condition to move on along the route that connects redox biomolecules and their application in single-molecule electronic devices. The final step is the demonstration of transistor functionality inherent in a redox metalloprotein. This has been, indeed, regarded as the *holy grail* of biomolecular electronics in search of which all the efforts of two decades of topical scientific research have been devoted. Demonstrating such a possibility has multiple impacts on our view of redox metalloproteins. The most evident is that connected with the development of bioelectronics as a *novel approach to electronics based on functional biomolecules.* As we will discuss further on, it is our opinion that this aspect is the weakest and the less relevant, since we can anticipate that an approach aiming at simply substituting solid-state electronics with bioelectronics does not appear very promising to date.

There is, however, a second aspect that is worthwhile, i.e., the realization of the role of functional biomolecules in general, and redox metalloproteins in particular, as self-contained functional units. This is an important concept independently of the application of redox metalloproteins to the implementation of electronic devices, since it points to the concept of functionality as a property *inherent to a particular application or context.* In this sense it is also relevant to the epistemological meaning of the word "functionality". There is, indeed, an increasing body of evidence that points to the role of molecular systems as the more appropriate sites of biological functionality, rather than the self-contained, single molecule (Alessandrini & Facci, 2011). Such a view can be applied also to the case of azurin, if we focus on its "physiological" functional activity that encompasses interaction with redox partners. Nevertheless, it appears clearly that azurin's *redox* functionality can be regarded as a property intrinsic to the molecule itself, and in such a sense it escapes the interpretation presented above for *biological* functional activity. Hence, we can recognize that the concept of functionality is not an absolute one; rather, it has to be specified in relation to the context where we want to apply it.

The demonstration of a single-molecule redox protein transistor stems from the use of the ECSTM experimental set-up in an innovative configuration and also provides a quantitative clue to the mechanism underlying electron transport through a single metalloprotein.

We have seen in section 4.5.3 that ECSTM of azurin performed as a function of substrate potential provides evidence for substrate potential dependent STM contrast. Those results are suggestive of a two-step tunneling mechanism but can be interpreted only at a qualitative level since constant current imaging prevents access to the tunneling current, due to a lack of knowledge about the β factor of the protein. The need for direct access to the tunneling current emerges, thus, as a critical aspect to go further in the analysis. Similarly to the case of quinone derivatives, one has to rely on tunneling spectroscopy if one is to fit theoretical trends to experimental data. In the case of metalloproteins, however, there exists a peculiar problem that makes things more complicated than in the aforementioned case. Although azurin is chemisorbed on the Au(111) surface by substantially covalent S-Au bonds, repeated scans, needed to find the molecules on the surface and possible tip—sample interactions during unavoidable, periodic adjustment of the tip—sample 3D mutual position required to maintain the correct experimental configuration, can easily damage the molecular conformation or even sweep out molecules from the scanned surface. This fact means that $I_t(V_S)$ spectroscopy is not easily achievable.

In order to circumvent these difficulties, a different experimental arrangement has turned out to be useful. Namely, by changing the role between tip and substrate, many of the aforementioned difficulties disappear. The idea is that of letting azurin chemisorb on the surface of an ECSTM gold tip. As such the gold tip has to be almost completely insulated and exposed to solution just its apex, where metalloproteins can chemisorb. Figure 4.19 shows a TEM image of such a biofunctionalized tip.

These tips show the typical electrochemical (CV) behavior expected from a nanoelectrode (almost negligible capacitance contribution), and reveal the retained redox activity of the chemisorbed molecules (Alessandrini et al., 2005).

With such a tip it is possible to solve a number of problems affecting the traditional spectroscopy-like ECSTM imaging. Indeed, one does not have to search for a molecule on a substrate surface by repeated scans, hence decreasing the wear effect due to the raster scan of the tip on the surface. Furthermore, this approach eliminates the need for tracking the molecule, an aspect that is usually very critical in tunneling spectroscopy. $X-y$ drift will be almost uninfluential on the measured I_t and z drift will be easily corrected before the beginning of each characteristic curve measurement.

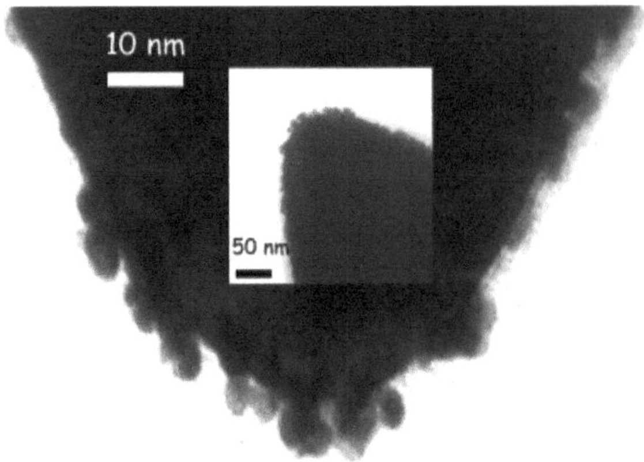

FIGURE 4.19

TEM image of an azurin-coated, wax-insulated Au tip. Several single molecules are visible. Tunneling current is likely to be picked up by the most protruding molecule due to the fast spatial decay of the tunneling current.

Adapted with permission from Alessandrini et al., Appl. Phys. Lett., *86, 133902, 2005. Copyright 2005, American Institute of Physics.*

The most relevant advantage of the protein-on-tip approach, however, is that it enables direct access to the tunneling current as a function of substrate potential. Indeed, here V_S and V_t flip their role and one will scan V_t and consider I_t as a function of it, rather than of V_S.

Experimentally, one has to engage a certain tunneling current set-point, and, after tunneling junction stabilization (e.g., thermal drift and piezo creep settling down), disable the feedback. Afterwards, one starts sweeping the tip voltage while measuring the tunneling current, i.e., $I_t(V_t)$.

The result of such measurements on an azurin-coated Au tip is the appearance of a marked resonance in a potential region close to the redox potential of the molecule. That feature is absent in both bare tips and Zn-azurin-coated tips, used as negative control measurements. It is worthwhile noting that the same molecule sustains repeated potential scans without showing a significant decrease in performance. Furthermore, the nature of the resonance, traced back to the involvement of electronic levels in a two-step tunneling mechanism as discussed below, is confirmed in some measurements which show the co-existing resonance and a faradaic wave ascribable to the molecules not on the apex of a tip that is not perfectly insulated (Alessandrini et al., 2005). Figure 4.20a reports these data.

Measuring the $I_t(V_t)$ (or, analogously, $I_t(V_s)$) characteristics at different V_{bias} (Figure 4.20b) it is possible to obtain a set of curves that are readily

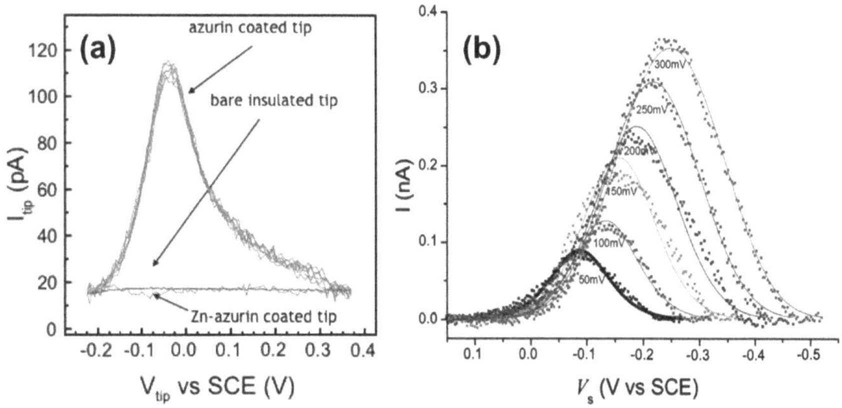

FIGURE 4.20

(a) $I_t(V_t)$ measured with Au tips. Only Cu-azurin yields current resonances, at variance with bare and Zn-azurin-coated tips; (b) set of $I_t(V_S)$ curves as a function of V_{bias}.

(a) Adapted with permission from Alessandrini et al., Appl. Phys. Lett., *86, 133902, 2005. Copyright 2005, American Institute of Physics. (b) From Alessandrini et al.,* Phys. Chem. Chem. Phys., *8, 4383–4397, 2006. Reproduced by permission of the PCCP Owner Societies.*

interpretable by two-step electron transfer theory (Friis et al., 1998). Fitting the theory (see section 4.5.2) to the experimental data provides the following results. Firstly, the theory that best fits the data appears to be two-step coherent electron tunneling with partial relaxation, even if an incoherent (fully relaxed) model cannot be completely figured out since it could be still at play if one hypothesizes a strong electronic coupling between azurin redox levels and those of the leads. Second, within that theoretical frame, it is possible to extract values for relevant parameters such as the redox level ε_0 and the reorganization energy λ. Whereas the first is consistent with the value for E_0' estimated from CV, the second appears much lower (0.13 eV) than that measured for the self-exchange reaction ($\approx$ 0.6 eV) (Alessandrini et al., 2006). This discrepancy is still a matter of debate but can be traced back to the ECSTM configuration itself, which forces the redox molecule between substrate and tip, two macroscopic objects, in a tiny gap from which most of the solvent is squeezed out. This fact can have a direct impact on the solvent-associated component of the reorganization energy, which is thus decreased.

It is now possible to regard the reported results from the standpoint of their relevance for applications. Indeed, the kind of functionality that emerges from the data resembles that of an electronic switch or molecular transistor. As a matter of fact, the behavior of a single azurin molecule in between two metal leads shows a striking analogy to that of a generic single-particle transistor. Generally speaking, the latter takes advantage of a capacitive gate coupled to single-particle levels and is able to tune them with respect to the Fermi levels of the two metal leads that play the role of source and drain and that are, as such, biased by an external voltage to drive current through the particle. Figure 4.21 depicts that situation. Only when particle levels are properly aligned to the Fermi levels of the leads can a sizable current flow through the particle.

The described situation is very similar to that encountered in the case of ECSTS of azurin. Even there, tuning the electronic redox levels of the molecule to the Fermi levels of the leads results in an enhanced tunneling current flow between electrically biased source and drain (tip and substrate, or vice versa). In this respect the analogy with a transistor functionality is straightforward. The main difference is the way used for achieving the proper tuning of the electronic level. Indeed, in the case of azurin, the gate is electrochemical in nature rather than capacitive as in the case of a single-particle

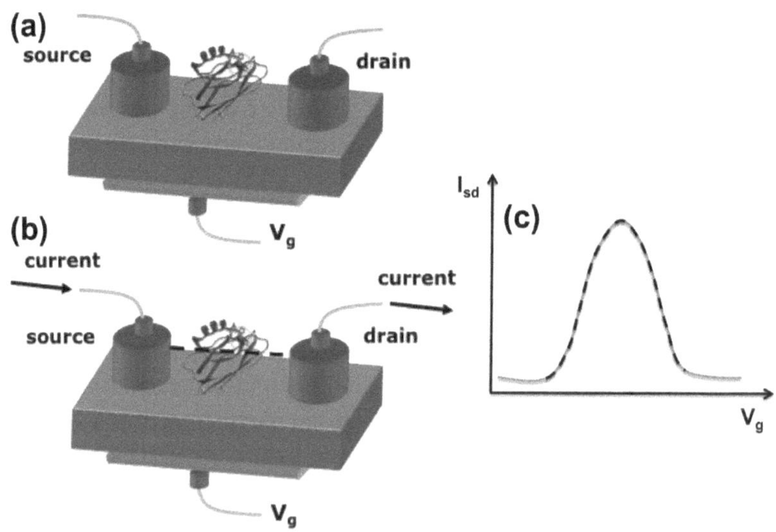

FIGURE 4.21

Scheme of a single-particle transistor. The role of the single particle in this representation is played by an azurin molecule. (a) Transistor is off; (b) transistor is on; (c) ideal plot of $I_{sd}(V_g)$. Note the analogy between the qualitative trend of the current in this case, with respect to that measured in experiments with azurin-coated tips in the ECSTM/STS configuration.

transistor. This means that it is somehow diffused and brought about by the supporting electrolyte, which enables potential control at tip and substrate following the action of the bipotentiostat.

Therefore, the behavior of azurin chemisorbed on an ECSTM tip apex is that of a single-protein transistor with an electrochemical gate.

The question arises now as to the applicability of these kinds of results to the implementation of real devices. We can preliminarily observe that the best on-off ratio one can estimate out of the reported data is still quite limited (< 500, in the best case), and not comparable with that of more conventional solid-state transistor devices (typical $I_{on}/I_{off} > 10{,}000$ for a MOSFET). This is, however, a remarkable result at room temperature, if one considers that similar figures are still hard to achieve even in recent graphene FET-type transistors (Xia et al., 2010). However, as far as transistor performance is concerned, the correct comparison is, according to our judgment, with standard solid-state devices rather than with single-particle ones (e.g., single electron transistor; SET) since that is the target of any innovative device that aims to assert itself as a relevant player in tomorrow's (nano) electronics.

Other problems arise in terms of durability and performance reliability of protein-based devices, not to mention the need for a wet environment to function. In this respect it is really hard, and frankly not very realistic, to imagine replacing the current solid-state electronic devices and fabrication strategy by any other technology involving bottom-up assembly starting from functional molecules, albeit promising, and even more stable than proteins. This belief is also driven by the consideration that the tremendous cost that the present fabrication facilities in the electronic components industry have reached in practice does not allow for any change of strategy without an even worldwide unaffordable change of perspective.

In order to find a space for the newborn biomolecule-based device concepts, we think that a safe enough solution could be that of seeking for niche applications. Among the possible scenarios, ultrasensitive biosensors down to the single-molecule level, for instance, could be more appealing and have realistic implementations. At any rate, the demonstration of an electronic switch or transistor functionality in a single metalloprotein represents a basic conceptual step towards the general goal of implementing an electrical control of biological reactions and phenomena, since it allows a detailed description of the fundamental transport mechanisms behind the electron flow via these molecules and the deep understanding thereof represents the basis for any further development in the abovementioned direction (Facci, 2009).

4.7 References

Adam, E. T. (1991). Copper protein structures. *Adv. Protein Chem., 42*, 145–197.

Aggett, P. J. (1985). Physiology and metabolism of essential trace elements: an outline. *Clin. Endocrinol. Metab., 14*, 513–543.

Albrecht, T., Guckian, A., Ulstrup, J., & Vos, J. G. (2005). Transistor-like behavior of transition metal complexes. *Nano Lett., 5*, 1451–1455.

Alessandrini, A., & Facci, P. (2005). AFM: a versatile tool in biophysics. *Meas. Sci. Technol., 16*, R65–R92.

Alessandrini, A., & Facci, P. (2007). Metalloprotein electronics. In S. E. Lyshevski (Ed.), *Nano and molecular electronics handbook*. Boca Raton, FL: CRC Press, 14-1–14-47.

Alessandrini, A., & Facci, P. (2011). Changes in single K^+ channel behavior induced by a lipid phase transition. *Communicative Integrat. Biol., 4*, 1–3.

Alessandrini, A., Gerunda, M., Canters, G. W., Verbeet, M. P., & Facci, P. (2003). Electron tunnelling through azurin is mediated by the active site Cu ion. *Chem. Phys. Lett., 376*, 625–630.

Alessandrini, A., Salerno, M., Frabboni, S., & Facci, P. (2005). Single-metalloprotein wet biotransistor. *Appl. Phys. Lett., 86*, 133902.

Alessandrini, A., Corni, S., & Facci, P. (2006). Unravelling single metalloprotein electron transfer by scanning probe techniques. *Phys. Chem. Chem. Phys., 8*, 4383–4397.

Alessandrini, A., Berti, L., Gazzadi, G. C., & Facci, P. (2008). Imparting chemical specificity to nanometer-spaced electrodes. *Nanotechnology, 19*, 355303.

Armstrong, F. A., Hill, H. A., & Walton, N. J. (1988). Direct electrochemistry of redox proteins. *Acc. Chem. Res., 21*, 407–413.

Avila, A., Gregory, B. W., Niki, K., & Cotton, T. M. (2000). An electrochemical approach to investigate gated electron transfer using a physiological model system: cytochrome c immobilized on carboxylic acid–terminated alkanethiol self assembled monolayers on gold electrodes. *J. Phys. Chem. B, 104*, 2759–2766.

Banci, L., & Assfalg, M. (2001). In A. Messerschmidt, et al. (Eds.), *Handbook of metalloproteins* (vol. 1). Chichester: Wiley.

Banci, L., Bertini, I., Rosato, A., & Varani, G. (1999). Mitochondrial cytochromes c: a comparative analysis. *J. Biol. Inorg. Chem., 4*, 824–837.

Bard, A. J., & Faulkner, L. R. (2001). *Electrochemical methods—fundamentals and applications* (2nd ed.). New York: John Wiley & Sons.

Bendall, D. S. (1996). Interprotein electron transfer. In D. S. Bendall (Ed.), *Protein electron transfer*. Oxford: BIOS Scientific.

Beratan, D. N., Onuchic, J. N., Betts, J. N., Bowler, B. E., & Gray, H. B. (1990). Electron-tunneling pathways in ruthenated proteins. *J. Am. Chem. Soc., 112*, 7915–7921.

Beratan, D. N., Betts, J. N., & Onuchic, J. N. (1991). Protein electron transfer rates set by the bridging secondary and tertiary structure. *Science, 252*, 1285–1288.

Beratan, D. N., Betts, J. N., & Onuchic, J. N. (1992). Tunneling pathway and redox-state-dependent electronic couplings at nearly fixed distance in electron-transfer proteins. *J. Phys. Chem., 96*, 2852–2855.

Binnig, G., & Rohrer, H. (1983). Scanning tunneling microscopy. *Surf. Sci., 126*, 236–244.

Binnig, G., Quate, C. F., & Gerber, C. (1986). Atomic force microscope. *Phys. Rev. Lett., 56*, 930–933.

Bortolotti, C. A., Borsari, M., Sola, M., Chertkova, R., Dolgikh, D., Kotlyar, A., & Facci, P. (2007). Orientation-dependent kinetics of heterogeneous electron transfer for cytochrome c immobilized on gold: Electrochemical determination and theoretical prediction. *J. Phys. Chem. C, 111*, 12100–12105.

Branden, C., & Tooze, J. (1999). *Introduction to protein structure* (2nd ed.). New York: Garland.

Brayer, G. D., & Murphy, M. E. P. (1996). In R. A. Scott & A. G. Mauk (Eds.), Cytochrome c: a multidisciplinary approach, *part II.* (p. 103). Sausalito, CA: University Science Books.

Brill, A. S. (1977). *Transition metals in biochemistry.* Berlin: Springer.

Bryce, C. F. A. (1979). SAM – semantics and misunderstandings. *Trends Biochem. Sci., 4*, N62–N63.

Cavalieri, R. R. (1997). Iodine metabolism and thyroid physiology: current concepts. *Thyroid, 7*, 177–181.

Chan, M. K., Mukund, S., Kletzin, A., Adams, M. W., & Rees, D. C. (1995). Structure of a hyperthermophilic tungstopterin enzyme, aldehyde ferredoxin oxidoreductase. *Science, 267*, 1463–1469.

Chi, Q., Zhang, J., Friis, E. P., Andersen, J. E. T., & Ulstrup, J. (1999). Electrochemistry of self-assembled monolayers of the blue copper protein *Pseudomonas aeruginosa* azurin on Au(111). *Electrochem. Commun., 1*, 91–96.

Clapham, D. E. (2007). Calcium signaling. *Cell, 131*, 1047–1058.

Cowan, J. A. (1997). Inorganic biochemistry: an introduction. New York: Wiley-VCH.

Devault, D. (1984). *Quantum mechanical tunnelling in biological systems.* Cambridge: Cambridge University Press.

Eady, R. R. (1988). The vanadium-containing nitrogenase of Azotobacter. *BioFactors, 1*, 111–116.

Eddowes, M. J., & Hill, H. A. O. (1977). Novel method for the investigation of the electrochemistry of metalloproteins: cytochrome c. *J. Chem. Soc. Chem. Commun., 21*, 771b–772.

Eddowes, M. J., & Hill, H. A. O. (1979a). Electrochemistry of cytochrome c. Comparison of the electron transfer at a surface-modified gold electrode with that to cytochrome oxidase. *J. Am. Chem. Soc., 101*, 7113–7114.

Eddowes, M. J., & Hill, H. A. O. (1979b). Electrochemistry of horse heart cytochrome c. *J. Am. Chem. Soc., 101*, 4461–4464.

Eddowes, M. J., & Hill, H. A. O. (1982). Factors influencing the electron-transfer rates of redox proteins. *Faraday Discuss. Chem. Soc., 74*, 331–341.

Elliott, D., Hamnett, A., Lettington, O. C., Hill, H. A. O., & Walton, N. J. (1986). The determination, by ellipsometry, of the mode of adsorption on gold of some promoters of redox-protein direct electrochemistry. *J. Electroanal. Chem., 202*, 303–314.

Facci, P. (2002). Single metalloprotein at work: towards a single-protein transistor. In T. Chakraborty, F. Peeters, & U. Sivan (Eds.), Nano-physics & bio-electronics: A new Odyssey. (pp. 323–339). Amsterdam: Elsevier.

Facci, P. (2009). Biomolecular electronics - an overview and future trends in biomolecular electronics. Nanotechnology thoughts leaders series www.azonano.com/article.aspx?ArticleID=2786.

Facci, P., Alliata, D., & Cannistraro, S. (2001). Potential-induced resonant tunneling through a redox metalloprotein investigated by electrochemical scanning probe microscopy. *Ultramicroscopy, 84*, 291–298.

Friis, E. P., Kharkats, Y. I., Kuznetsov, A. M., & Ulstrup, J. (1998). In situ scanning tunneling microscopy of a redox molecule as a vibrationally coherent electronic three-level process. *J. Phys. Chem. A, 102*, 7851–7859.

Gerischer, H. (1960). Kinetics of oxidation-reduction reactions on metals and semiconductors. I. General remarks on the electron transition between a solid body and a reduction-oxidation electrolyte. *Z. Phys. Chem. NF, 26*, 223–270.

Gerunda, M., Bortolotti, C. A., Alessandrini, A., Sola, M., Battistuzzi, G., & Facci, P. (2004). Grabbing yeast iso-1-cytochrome c by Cys102: an effective approach for the assembly of functionally active metalloprotein carpets. *Langmuir, 20*, 8812–8816.

Gray, H. B., & Ellis, W. (1994). Electron transfer in biology. In I. Bertini, et al. (Eds.), *Electron transport metalloproteins in bioinorganic chemistry*. Sausalito, CA: University Science Books.

HARLEM software, Kurnikov, IV, http://www.kurnikov.org/.

Harrison, P. M. (1985). *Metalloproteins Parts I and II*. New York: Academic Press.

Holm, R. H., Kennepohl, P., & Solomon, E. I. (1996). Structural and functional aspects of metal sites in biology. *Chem. Rev., 96*, 2239–2314.

Jacq, J. (1971). Schema carre. Establissement et discussion de l'equation generale de la courbe intensite-potentiel en regime stationnaire et diffusion convective. *J. Electroanal. Chem., 29*, 149–180.

Jordan, F., & Patel, M. S. (2004). *Thiamine: catalytic mechanisms in normal and disease states*. New York: Marcel Dekker.

Khoshtariya, D. E., Wei, J., Liu, H., Yue, H., & Waldeck, D. H. (2003). Charge-transfer mechanism for cytochrome c adsorbed on nanometer thick films. Distinguishing frictional control from conformational gating. *J. Am. Chem. Soc., 125*, 7704–7714.

Koppenol, W. H., & Margoliash, E. (1982). The asymmetric distribution of charges on the surface of horse cytochrome c. Functional implications. *J. Biol. Chem., 257*, 4426–4437.

Kuznetsov, A. M., & Schmickler, W. (2002). Mediated electron exchange between an electrode and the tip of a scanning tunneling microscope - a stochastic approach. *Chem. Phys., 282*, 371–377.

Landrum, H. L., Salmon, R. T., & Hawkridge, F. M. (1977). A surface-modified gold minigrid electrode which heterogeneously reduces spinach ferredoxin. *J. Am. Chem. Soc., 99*, 3154–3158.

Lane, T. W., & Morel, F. M. (2000). A biological function for cadmium in marine diatoms. *Proc. Natl. Acad. Sci. USA, 97*, 4627–4631.

Lange, C., & Hunte, C. (2002). Crystal structure of the yeast cytochrome bc(1) complex with its bound substrate cytochrome c. *Proc. Natl. Acad. Sci. USA, 99*, 2800–2805.

Laviron, E. (1984). Electrochemical reactions with protonations at equilibrium. Part X. The kinetics of the p-benzoquinone/hydroquinone couple on a platinum electrode. *J. Electroanal. Chem., 164*, 213–227.

Lide, D. R. (Ed.). (1993). *CRC handbook of chemistry and physics*. Boca Raton, FL: CRC Press.

Meyer, J. (2008). Iron-sulfur protein folds, iron-sulfur chemistry, and evolution. *J. Biol. Inorg. Chem., 13*, 157–170.

Migliore, A., Corni, S., Di Felice, R., & Molinari, E. (2006). First-principles density-functional theory calculations of electron-transfer rates in azurin dimers. *J. Chem. Phys.* 124, 064501.

Moser, C. C., & Dutton, P. L. (1992). Engineering protein structure for electron transfer function in photosynthetic reaction centers. *Biochim. Biophys. Acta, 1101*, 171–176.

Nahir, T. M., & Bowden, E. F. (1996). The distribution of standard rate constants for electron transfer between thiol-modified gold electrodes and adsorbed cytochrome c. *J. Electroanal. Chem., 410*, 9–13.

Nar, H., Messerschmidt, A., Van De Kamp, M., Canters, G. W., & Huber, R. (1991). Crystal structure analysis of oxidized *Pseudomonas aeruginosa* azurin at pH 5.5 and pH 9.0. A pH-induced conformational transition involves a peptide bond flip. *J. Mol. Biol., 221*, 765–772.

Nic, M., Jirat, J., & Kosata, B. (Eds.). (2006). *Quinones. IUPAC compendium of chemical terminology, online ed.*

Niki, I., Yokokura, H., Sudo, T., Kato, M., & Hidaka, H. (1996). Ca2+ signaling and intracellular Ca2+ binding proteins. *J. Biochem., 120*, 685–698.

Page, C. C., Moser, C. C., Chen, X., & Dutton, P. L. (1999). Natural engineering principles of electron tunnelling in biological oxidation-reduction. *Nature, 402*, 47–52.

Park, W., & Hong, H. G. (2006). Determination of reorganization energy from the temperature dependence of electron transfer rate constant for hydroquinone-tethered self-assembled monolayers (SAMs). *Bull. Korean Chem. Soc., 27*, 381–385.

Park, W., Ahmed, J., & Kim, S. (2009). Heterogeneous electron-transfer kinetics for PQQ covalently attached to aminoalkanethiol monolayers on gold. *Colloids Surf. B, 68*, 120–124.

Pascher, T., Chesick, J. P., Winkler, J. R., & Gray, H. B. (1996). Protein folding triggered by electron transfer. *Science, 271*, 1558–1560.

Petrangolini, P., Alessandrini, A., Berti, L., & Facci, P. (2010). An electrochemical scanning tunneling microscopy study of 2-(6-mercaptoalkyl)hydroquinone molecules on Au(111). *J. Am. Chem. Soc., 132*, 7445–7453.

Petrangolini, P., Alessandrini, A., Navacchia, M. L., Capobianco, M. L., & Facci, P. (2011). Electron transport properties of single-molecule-bearing multiple redox levels studied by EC-STM/STS. *J. Phys. Chem. C, 115*, 19971–19978.

Salomon, A., Cahen, D., Lindsay, S., Tomfohr, J., Engelkes, V. B., & Frisbie, C. D. (2003). Comparison of electronic transport measurements on organic molecules. *Adv. Mater., 15*, 1881–1890.

Saraste, M. (1999). Oxidative phosphorylation at the fin de siècle. *Science, 283*, 1488–1493.

Sauke, D. J., Metzler, D. E., & Metzler, C. M. (2001). *Biochemistry: the chemical reactions of living cells* (2nd ed.). San Diego, CA: Harcourt/Academic Press.

Schmickler, W. (1993). Investigation of electrochemical electron transfer reactions with a scanning tunneling microscope: a theoretical study. *Surf. Sci., 295*, 43–56.

Schmickler, W., & Tao, N. J. (1997). Measuring the inverted region of an electron transfer reaction with a scanning tunneling microscope. *Electrochim. Acta, 42*, 2809–2815.

Schmickler, W., & Widrig, C. (1992). The investigation of redox reactions with a scanning tunneling microscope. Experimental and theoretical aspects. *J. Electroanal. Chem., 336*, 213–221.

Schnyder, B., Kotz, R., Alliata, D., & Facci, P. (2002). Comparison of the self-chemisorption of azurin on gold and on functionalized oxide surfaces. *Surf. Interface Anal., 34*, 40–44.

Siegenthaler, H. (1995). STM in electrochemistry. In R. Wiesendanger, & H.-J. Güntherodt (Eds.), *Scanning tunneling microscopy II* (2nd ed.). Berlin: Springer.

Vetter, K. J. (1952). *Elektrochem., 56*, 79.

Weber, K., & Creager, S. E. (1994). Voltammetry of redox-active groups irreversibly adsorbed onto electrodes. Treatment using the Marcus relation between rate and overpotential. *Anal. Chem., 66*, 3164–3172.

Xia, F., Farmer, D. B., Lin, Y.-M., & Avouris, P. (2010). Graphene field-effect transistors with high on/off current ratio and large transport band gap at room temperature. *Nano Lett., 10*, 715–718.

Yeh, P., & Kuwana, T. (1977). Reversible electrode reaction of cytochrome C. *Chem. Lett., 6*, 1145–1148.

Zhang, J., Chi, Q., Kuznetsov, A. M., Hansen, A. G., Wackerbath, H., Christensen, H. E. M., Andersen, J. E. T., & Ulstrup, J. (2002). Electronic properties of functional biomolecules at metal/aqueous solution interfaces. *J. Phys. Chem. B, 106*, 1131–1152.

Zhang, J., Chi, Q., Albrecht, T., Kuznetsov, A. M., Grubb, M., Hansen, A. G., Wackerbarth, H., Welinder, A. C., & Ulstrup, J. (2005). Electrochemistry and bioelectrochemistry towards the single-molecule level: theoretical notions and systems. *Electrochim. Acta, 50*, 3143–3159.

CHAPTER

Electrochemistry can Drive Molecular Conformation

5

CHAPTER OUTLINE

5.0 Direct electrochemical control of protein conformation at an electrode surface

Direct electrochemistry is not limited to affecting redox molecular moieties. Indeed, the action of an electrochemical field at an interface between, for instance, an electrode and a liquid containing an electrolyte can influence the conformation of thereby immobilized biomolecules to the extent of inducing a transition from one state to another and, possibly, also causing a change in the corresponding functional state of the molecule (or "global-state"; Jackson, 2006).

Such a result can be achieved by the action of an applied electric field on charges and charge distributions brought about by biomolecules in physiological or artificial environments. In fact, the presence of charged amino acid side chains on the one hand and of effective dipolar moments on the

Biomolecular Electronics. http://dx.doi.org/10.1016/B978-1-4557-3142-8.00005-4

other represent the main sources of free charges and dipoles in biomolecules, respectively. Among amino acid side chains, in physiological pH conditions primary amines of both lysines and arginines are virtually completely ionized (as indicated by their pK_a values, typically above 10; Bashford, 2004). Dipole moments, for example, arise regularly from the unsaturation of carboxyl and amino termini in alpha helices, being unable to form intramolecular H-bonds with nearby residues (e.g. between the nth and $n+4^{th}$ residue in the case of an alpha helix).

Whereas examples of applied electric fields interacting with biomolecules and biosystems are numerous and widely exploited even in technology (e.g., electrophoresis, electroporation, cell sorting, etc., to cite just a few), we are here mainly interested in discussing paradigmatic cases of the effect of an external electric field on the conformation of surface immobilized biomolecules and in evaluating the functional consequence that the action of the electric field shows on the functional behavior of the proteins at issue.

In the next sections we will see two examples involving very different proteins, i.e., a voltage-gated, potassium ion channel and antibodies of the IgG type. In both cases the application of an electric field between a gold electrode and the bulk of the solution is able to affect the conformation of surface immobilized proteins up to switching their functional state in a predictable way. Since such an electric field will have a direction perpendicular to the electrode surface, its effective action will deploy along that same direction, affecting those molecular domains that have degrees of freedom along that axis. This fact imposes strict requirements onto the way biomolecules are arranged on the electrode surface, namely onto their orientation and on the kind of chemical/physical interaction exploited to attach them to the surface.

5.1 Direct electrical modulation of the open/closed state of a voltage-gated potassium ion channel

We have learnt much information about voltage-gated K^+ channels in section 3.6.1; particularly, we have seen how it is possible to measure fine details of the functional behavior and related conformational features of these channels by electrophysiological techniques such as patch and voltage clamp (Hille, 2001). However, these approaches, albeit information-rich, provide only indirect evidence on the conformational modifications occurring upon gating a channel by a modulation in the transmembrane potential. A genuine, single-molecule, approach would be that enabling us to see the conformational modifications associated with a change in the transmembrane potential. Toward such an ambitious goal one has typically to exploit the potential of the nanoscience approach to biophysics and molecular biology, that is, in this case, to take advantage of experimental tools such as scanning probe microscopes. The relevance of scanning probe microscopy for nanobiophysics is widely recognized and the interested reader can find a ready reference in any of the numerous reviews that have been published on this topic (e.g., Alessandrini & Facci, 2005).

There are, however, several features to control in order to design and assemble an experiment suitable to achieve the desired goal. First of all, one has to choose the right molecule; we opt for a voltage-gated potassium ion channel from a green plant, *Arabidopsis thaliana*, that possesses the interesting feature of being an inward rectifier (see section 3.6.1), hence moving its voltage sensors towards the inner part of the membrane upon gating. This is an important aspect since it can be considered as a necessary condition for the visualization of the effect of the gating at a conformational level when the inner side of the membrane is to be inspected by an AFM.

Second, it is necessary to deposit the sample on a flat enough electrode surface to allow for an adequate resolution in AFM imaging; in other words, electrode surface roughness has to be low enough to allow for detecting submolecular details.

Third, the sample has to be properly attached to the substrate in order to sustain tip scans but has to retain enough mobility to be able to respond to changes in electric field.

Fourth, a suitable method has to be devised for excising a membrane patch and depositing it onto the electrode surface.

Fifth, a suitable approach for AFM probing of the different conformations corresponding to the open and closed states has to be implemented. This has to be compatible with imparting transmembrane potential values that correspond to open and closed channel states, respectively, in electrophysiological measurements. Whereas in the latter case the goal of imposing the desired potential drop across the membrane is guaranteed by the insulating behavior of a membrane patch (as revealed by the "gigaseal" that is established at the beginning of the experiment; see, e.g., Suchyna et al., 2009), the situation is quite different in the case of a membrane patch adsorbed onto the surface of an electrode. In this case, a three-electrode electrochemical cell has to be implemented (see Chapter 2) where the substrate plays the role of the working electrode (WE), and one has to play with the ionic strength of the solution in order to match the right Debye length (see section 2.2) that ensures the desired potential drop across the membrane.

We will see in detail the implementation of the relevant strategies suitable for addressing these points in the following sections. As a general observation, attaining the proper concurrency of all the aforementioned experimental details is usually not trivial and never a matter of few days, making the experiment we are going to describe a nice piece of interdisciplinary work that exemplifies very well the typical biophysical approach to biological problems.

5.1.1 The molecule

As already said, the choice for the present study regards the voltage-gated channel KAT1 from *Arabidopsis thaliana*, an inward rectifier potassium channel that is very well characterized from an electrophysiological standpoint (Hoshi, 1995; Schachtman et al., 1992) and whose structure, albeit not yet solved, is believed to resemble closely that of KvAP. The topology of plant inward rectifier channels is indeed similar to that of depolarization-activated Shaker-like K1 channels, consisting of four subunits, each made up of six transmembrane domains, a pore loop between the fifth and the sixth transmembrane regions and a voltage-sensing domain. Work by Latorre et al. (Latorre et al., 2003) has demonstrated that, when the Kat1 channel opens, the movement of the S4 segment charges is inward.

Therefore, because of its voltage-induced conformational modifications, this protein appears as a good candidate to be investigated in SFM experiments carried out on membrane patches deposited onto a substrate in inside-out configuration.

In order to deal with a convenient manipulation of biological membranes hosting the ion channel at issue, heterologous expression of Kat1 channels has been chosen by microinjecting cRNA into oocytes obtained surgically from large females of *Xenopus laevis* (Picco et al., 2004). These oocytes are gigantic cells (a fraction of a millimeter in diameter) allowing for single-cell manipulation under the optical microscope. Typically, 45 nl of a highly concentrated cRNA (2.4 mg/ml) have been injected into each cell to achieve a high expression level, thus increasing the probability of finding channels in SFM high-resolution scans. Incorporation of functional channels can be routinely tested using the two-electrode voltage-clamp technique, measuring the whole cell Kat1 current with a homemade voltage clamp amplifier.

Indeed, the expression level of Kat1 in *Xenopus* oocytes reaches a maximum 48 h after injection. Traces reported in Figure 5.1a show macroscopic currents recorded from an injected oocyte and have been elicited by decreasing the voltage from +20 mV to −160 mV in −10 mV steps; the

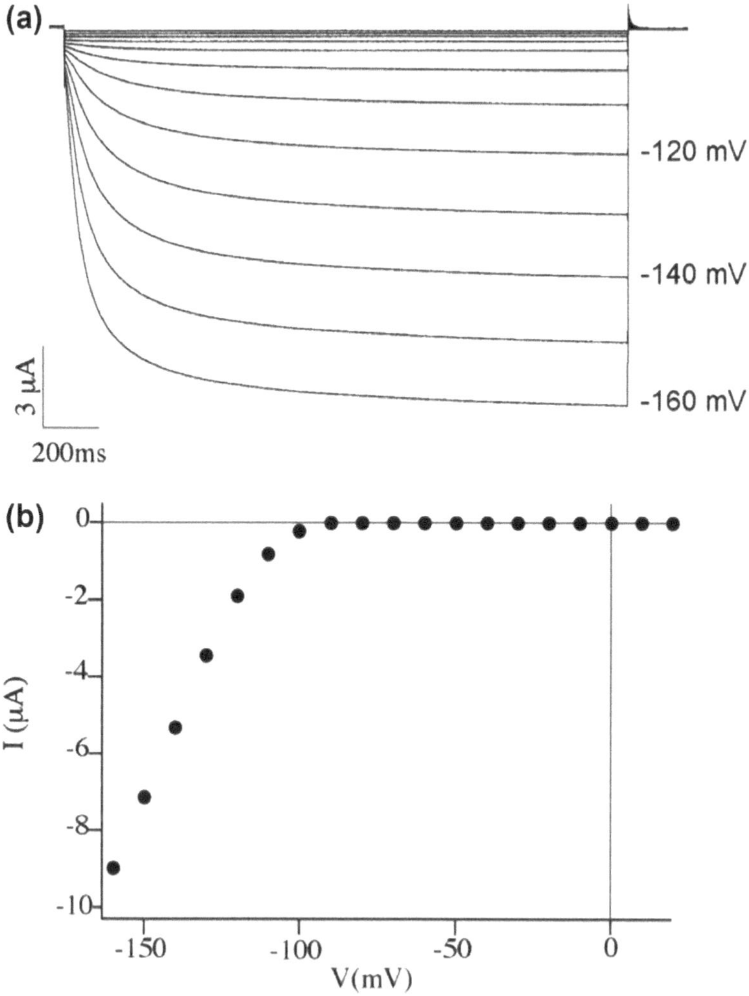

FIGURE 5.1

(a) Whole-cell current traces from a Kat1-expressing oocyte recorded from −160 to +20 mV. The external solution contained (in mM): KCl 100, $CaCl_2$ 1, $MgCl_2$ 2, MES/Tris 10; pH 5.6. The voltage clamp episode lasted 2 s. (b) I-V curve relative to the above traces. Plotted values are the average of the last 60 ms for each trace.

Reprinted with permission of John Wiley and Sons from Alessandrini et al., Microsc. Res. Techniq., *71, 274–278, 2008.*

relative current–voltage characteristic is reported in Figure 5.1b, where the average value of the last 60 ms of each current trace recording is plotted against the applied voltage.

These data confirm the presence of functional Kat1 channels and, given an oocyte diameter estimated at 1 mm, and the conductance of the Kat1 channel being typically 7 pS in symmetrical 100 mM K^+, they can be used to assess the density of three active channels per square micrometer.

Endogenous currents (i.e., those due to channels naturally belonging to the cell) have been previously observed and described in *Xenopus* oocytes, and also in these experiments an inward current, probably carried by Cl_2 ions (Dascal, 1987), sometimes disturbed recordings at membrane potential values more negative than -140 mV. To circumvent this problem, we decided to discard batches of oocytes showing large endogenous currents and decided to fix the potential for SFM imaging at -120 mV, a value very close to the half activation potential of Kat1 channels, while limiting possible potential-induced changes in endogenous channels.

It is to be noted that a figure of three per square micrometer for the expression level of functional channels, while remarkable for electrophysiology, appears quite challenging for the kind of experiment we are going to face, that is, finding the channels on the surface of a membrane patch with the scanning tip of an SFM!

5.1.2 Substrate and surface immobilization strategy

In order to succeed in applying an electric field across a supported membrane patch, while imaging it with a scanning force microscope, a mandatory step is to deposit a membrane patch on the surface of a polarizable electrode that shows a roughness value small enough to enable high-resolution (of the order of 1 nanometer) imaging. Our choice is that the substrate plays the role of a working electrode of a three-electrode electrochemical cell (see section 2.5), as we will discuss later on. One of the most convenient materials for making such electrodes is gold, as it can be readily deposited as a thin film on the surface of various supports (glass, SiO_2, mica, etc.) routinely used in scanning probe microscopy experiments. Thin ($\approx$ 100 nm) gold films can be typically deposited on those substrates by ion sputtering, e-gun, or thermal evaporation and a number of approaches have been developed to decrease their surface roughness (more precisely, the root-mean-square value of the roughness) down to values enabling high-resolution imaging. Among these techniques and depending on the support to be used, thermal annealing in vacuum, flame annealing, or template stripping from a mica surface can be used (or a combination of the first two approaches). In the case at issue, we made use of the first method, first evaporating a 150 nm-thick Au film on a freshly cleaved mica sheet and then thermally annealing it in vacuum (4 h at 450 °C). The result is a gold surface displaying a typical rms roughness of 0.4 nm.

Another, perhaps more relevant, feature makes the described choice the most attractive by far from an experimentalist's standpoint. A clean gold surface is indeed readily prone to functionalization by straightforward thiol-based chemistry. Indeed, once a flat enough electrode surface has been achieved, it is necessary to deposit over it a membrane patch that represents the sample to be imaged. It is important to realize that scanning force techniques require the sample be sufficiently immobilized on the substrate to prevent the scanning tip and the subsequent dragging effects from inducing sample movements at the surface. These would completely impair the resolution, making molecular and submolecular recognition unfeasible. The sample immobilization strategy, however, has to avoid a too strong attachment of the biological specimen to the substrate, which would make any gating-induced conformational modifications impossible. In the search for such an optimal balance between the two opposing requirements of firm substrate adhesion and preservation of protein conformational modifications, one can effectively opt for a chemistry that targets primary amines (present typically in surface-exposed lysines and arginines in surface-bound proteins and in some phospholipids; e.g., those characterized by PE headgroups). This approach has the advantage of preserving the freedom of movement of those proteins and/or protein motifs that are buried below the membrane surface, while taking advantage of those primary amines that are exposed off the surface (and that belong to molecules that are not our current experimental target). In practice, the mica-supported gold film was incubated in 1 mg/ml 2-mercaptoethylamine (2-MEA) water solution for 5 min, followed by glutaric dialdheyde (GD) incubation (1% water solution) for 10 min (Facci et al., 2002). This treatment makes the substrate surface adhesive for primary amine moieties (e.g., lysines) and is useful for immobilizing membrane patches by anchoring proteins at the surface.

5.1.3 Strategy for excising membrane patches and adsorbing them onto the electrode surface

After having ascertained that overexpression of the Kat1 channels has taken place in the microinjected oocytes, further processing is needed to produce the membrane patches to be imaged. Generally speaking it is possible to

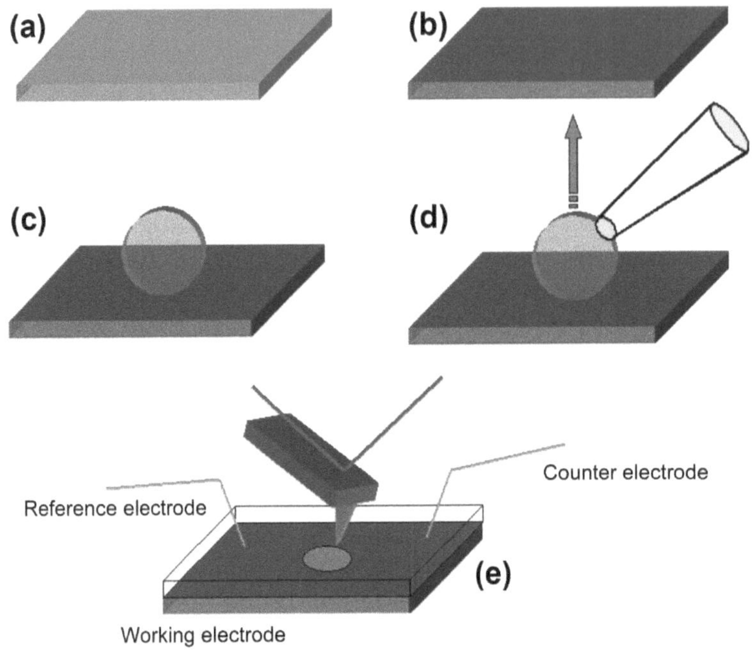

FIGURE 5.2

Steps of sample preparation. (a) A gold film is evaporated on a mica sheet; (b) the substrate is derivatized with 2-MEA + GD; (c) a microinjected oocyte is placed on the derivatized gold surface; (d) after some minutes it is removed with the help of a pipette; (e) a patch remains on the surface and SFM with electrochemical control is performed on it.

Reprinted with permission of John Wiley and Sons from Alessandrini et al., Microsc. Res. Techniq., *71, 274–278, 2008.*

follow well-described procedures in biological chemistry to prepare membrane patch suspensions following various steps that include cell disruption (usually by a French press), followed by ultracentrifugation in sucrose gradient and recovery of the fractions containing the desired membranes. Such an approach, albeit quite standardized and able to provide a reasonable amount of membrane patches, suffers one main trouble, which is the indeterminacy of patch orientation once adsorbed onto the functionalized gold substrate. There is, indeed, virtually no way to predict whether the deposited patch is in an inside-out or outside-out configuration, in electrophysiologist's terminology; the only possible approach to reveal patch orientation is to use labeled antibodies specific for some markers known to be present *only* in the inner (outer) side of the membrane.

A more straightforward, albeit a bit tricky, solution is that of manipulating a single oocyte at a time, sitting it on the functionalized surface in such a way as to be sure that the membrane patches that one eventually gets at the surface are all in the inside-out configuration. To achieve this goal, the macroscopic size of these oocytes helps. Indeed, after some further dissection steps (performed with a couple of tweezers under an optical microscope) that end with the removal of the *vitelline membrane* that wraps each oocyte, leaving the plasma membrane exposed, one is ready to adsorb the cell onto the substrate. By the help of a suction pipette, the oocyte can be sucked and gently put on the functionalized gold surface for typically 5 min; afterwards, the pipette is retracted and the oocyte removed. The oocyte footprint on the functionalized surface, a quite large membrane patch (not always uniform but sometimes showing fringes) is thus excised and attached to the surface. As the oocyte membrane is excised, unfortunately, the cytoplasmic content tends to exit the envelope, rapidly diffusing in the incubation chamber. This fact can cause the drawback of allowing a large amount of cytoplasmic material to adsorb onto the functionalized surface, disturbing further SFM imaging. Therefore, to minimize such a phenomenon, the oocyte has to be rapidly withdrawn and the incubation chamber thoroughly perfused with pure buffer. Figure 5.2 shows the main steps of the described procedure.

5.1.4 Measuring chamber implementation and approach for imparting a transmembrane potential drop

The measuring chamber is indeed a three-electrode electrochemical cell where the role of the working electrode is played by a gold substrate, and where a counter and a reference electrode are also present. These three electrodes are then connected to an external potentiostat capable of driving the potential of the substrate with respect to the solution. In the specific implementation of scanning force microscopy with electrochemical control, since measuring cells are usually quite small, one often prefers to go for a quasi-reference rather then a true reference electrode (e.g., Ag/AgCl, Hg/HgCl, etc.). This is, for example, a Ag wire whose potential against a true reference has to be checked before and after the measurements in the imaging buffer (to account for potential drift due to possible chemical modification of the Ag wire surface that does not behave ideally as a reference electrode). The counter or sacrificial electrode is usually made of platinum. Figure 5.2e shows a schematic view of a possible measuring set-up.

The buffer used in the imaging experiments was chosen, on the basis of electrophysiology experience, to be Barth's solution, with the following composition in mM: NaCl 88, KCl 1, $MgCl_2$ 0.82, $Ca(NO_3)_2$ 0.33, $CaCl_2$ 0.41, $NaHCO_3$ 2.4, Tris-HCl 5; pH 7.4. Barth's solution thus plays the double role of physiological buffer and supporting electrolyte for the electrochemical cell.

The described configuration enables one, in principle, to impart a potential drop between the metal substrate and the bulk of solution, and hence a voltage difference across the two sides of a membrane that is adsorbed onto the electrode surface, facing the solution. This situation is made possible by the value of the Debye length in the imaging solution (some nanometers), which allows a sizable potential drop off the electrode surface be localized across the sample membrane. We chose to operate at substrate potential values (i.e., at potential differences between the two sides of the membrane) indicated by electrophysiological measurements: in particular, +120 mV and −20 mV, corresponding to open and closed channel states, respectively (note that the sign of the potential is opposite here with respect to the usual electrophysiology standard).

5.1.5 Imaging strategy and results

The goal of retrieving information on possible conformational modifications of ion channels induced by gating them with an electric field is here addressed by means of SFM imaging. Indeed, this technique possesses the lateral resolution required for that task (Alessandrini & Facci, 2005). Nonetheless, instrumental resolution is only one of the ingredients needed to guarantee successful imaging. The already discussed substrate and sample roughness, stability, limited contamination by other molecules, etc., are concomitant requirements that have to be met to hope for success.

In the specific case, moreover, the aim is to appreciate tiny differences in ion channel conformation that occur upon functional state change (open/closed). The chosen imaging mode therefore has to be very uninvasive in order to allow for optimal resolution and minimum probe-sample interaction (this last aspect being absolutely needed when investigating soft biological samples). The imaging mode of choice for such a goal is the alternating contact mode (Alessandrini & Facci, 2005). We have chosen a magnetic implementation in order to minimize sample disturbance (Han et al., 1996). Briefly, the cantilever is here coated with a magneto-restrictive material that responds to a magnetic field generated by a proximal coil; contractions and elongations of the magnetic coating coupled to the back of the cantilever make this oscillate at a frequency close to its natural one. As in any alternating contact mode, imaging is achieved by controlling the percentage damping of the pre-set free oscillation amplitude as the probe contacts the sample, and exploiting the voltage applied to the vertical piezoelectric actuator needed to control the probe's vertical position, in order to build up a 2D topographical map of the scanned sample. Details and critical appraisals of this and other SFM imaging techniques can be found in the abundant literature reviewing SFM bioimaging (see, e.g., Alessandrini & Facci, 2005).

Highlighting by SPM subtle differences in Kat1 conformation as a function of its functional state requires that one operates in a differential way, acquiring topographic images of a given membrane area as a function of substrate potential, hence of the transmembrane potential that is imposed on the sample. Indeed, it is only from the fine comparison between the topographic information retrieved by imaging a given portion of sample at different substrate potential values that it is possible to appreciate differences connected to different conformational states the ion channel can assume. Unfortunately, it is not possible to say a priori if a membrane region contains the right molecule, since one observes just bumps where proteins are located. Therefore, one can get a final answer on the presence of the desired "functional" molecules just if one can outline localized morphological differences upon comparing images acquired at different substrate potentials.

Of course the first rough steps in the zoom-in process down to the molecular level can be taken without the complication of performing differential imaging, since they are aimed just at focusing on a membrane patch putatively suitable to be imaged at higher resolution and at different values of substrate potential.

Figure 5.3 shows, as an example, a membrane patch found on a functionalized mica surface. It is interesting to note that the region surrounding the patch, in the shape of a leaf, is decorated with bumps ascribable to soluble proteins coming from the oocyte cytosol, which have been adsorbed onto the reactive mica surface upon membrane excision. Their presence partially influences the membrane patch apparent height, which, nonetheless, appears consistent with what is expected for a cytoplasmic membrane patch.

Once a suitable patch has been found, imaging with direct electrochemical control of substrate potential value allows one to seek for zones that have shown morphological modifications in response to changes in the substrate (i.e., transmembrane potential). Operating in the proper potential interval, along with blocking the intrinsic ion currents with proper solution compositions, provides reasonable confidence in the possibility of affecting the state of just Kat1 channels. As already mentioned, maximum overexpression results in about three active channels per square micrometer, a figure that makes finding a shape-changing spot on the sample surface extremely unlikely. As a consequence, one needs to acquire several pairs of images

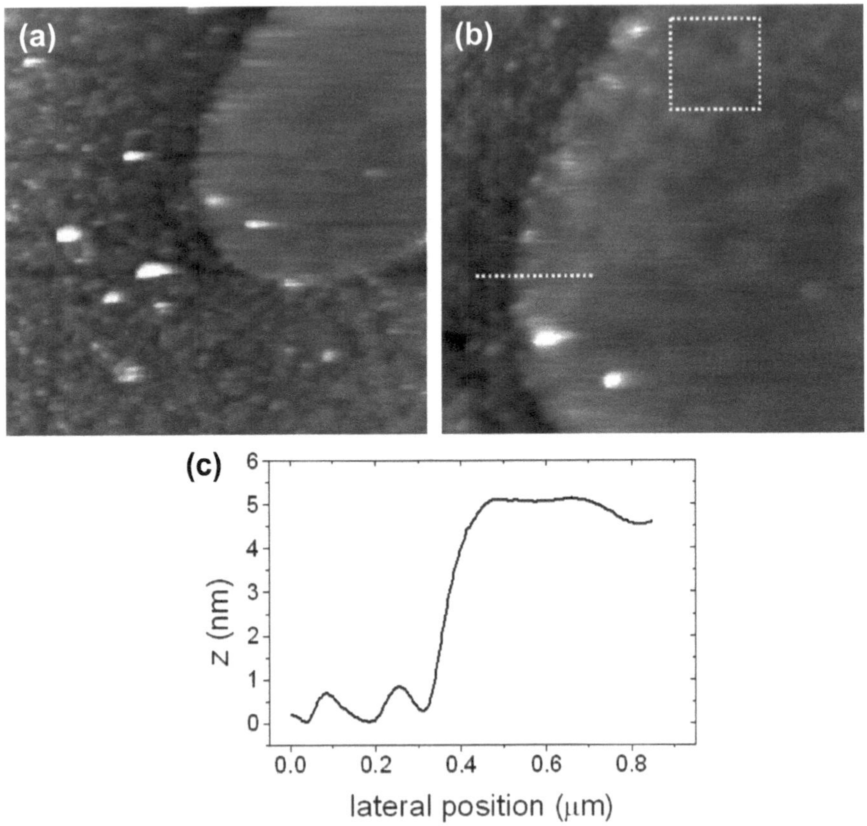

FIGURE 5.3

(a, b) Low-resolution topographic images of an oocyte membrane patch on functionalized gold on mica (image sizes: a, 3.9×3.9 μm^2; b, 2.3×2.3 μm^2). (c) Height profile along the dashed line in (b) that confirms the expected value of membrane thickness.

Reprinted with permission of John Wiley and Sons from Alessandrini et al., Microsc. Res. Techniq., *71, 274–278, 2008.*

(at potentials corresponding to open and closed states) to sort out those that appear significantly different. In Figure 5.4 an example of positive case is shown. The conformation of the feature that appears as a bump at a potential corresponding to the Kat1 closed state changes drastically on switching the potential to a value typical of the channel open state. Furthermore, the novel apparent feature is somehow shaped like a cross. It is surprising how such a form bears a close resemblance to the tetrameric shape of ion channels of the Shaker superfamily, as revealed by X-ray protein crystallography

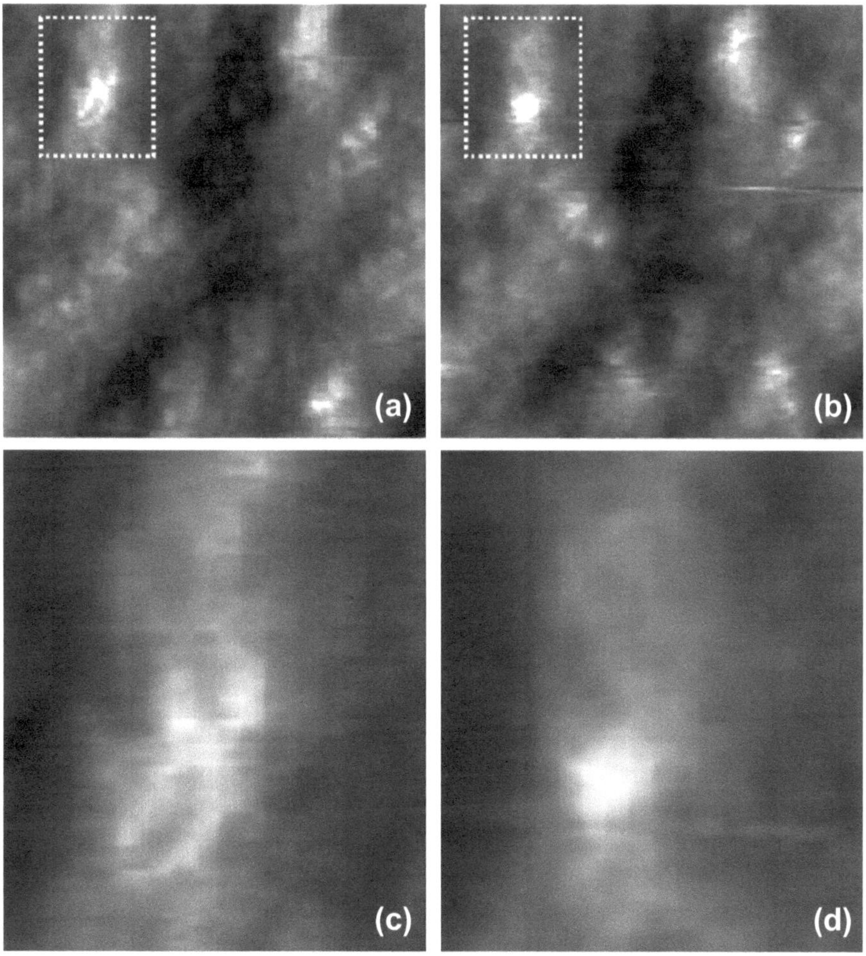

FIGURE 5.4

Higher-resolution images acquired at different potentials (−120 mV (a, c); +20 mV (b, d)) corresponding to open and closed channel states. Image size: a and b, 400×400 nm^2; c and d, 110×127 nm^2. Images (a) and (b) were acquired on the area outlined in Figure 5.3(b); images (c) and (d) were acquired on the area outlined in (a).

Reprinted with permission of John Wiley and Sons from Alessandrini et al., Microsc. Res. Techniq., *71, 274–278, 2008.*

(Jiang et al., 2003). The particular shape that appears makes one easily believe that these images are really showing a single ion channel having changed its conformation upon voltage gating.

Furthermore, other pieces of evidence corroborate this statement. Indeed, the apparent behavior is consistent with the choice of the particular channel for this experiment. Kat1 is known to be an inward rectifier; as such it is believed to move its voltage-sensitive domains towards the inner part of the plasma membrane upon opening in response to a change in transmembrane voltage. In this experiment, it is the inner side of the membrane that faces off the substrate and that is imaged by AFM, and the results appear to confirm this hypothesis.

It is also worth noticing that part of the channel structure appears to protrude off the membrane surface since this can contribute to shedding light on the molecular mechanisms associated with the event of voltage gating, helping to distinguish among different models for that phenomenon (see section 3.6.1).

Interestingly, the overall size of the shape-changing features well exceeds that of a single tetrameric channel as derived from the X-ray structures of similar molecules. This fact is only partially ascribable to phenomena such as tip–sample convolution that typically affect AFM images (Alessandrini & Facci, 2005). Indeed, the overall apparent size spans about 30 nm, a figure that cannot be accounted for by that kind of artifact. Rather, it appears that a sizable lipid region around the shape-changing features is affected in its structure by the effect of voltage gating. This evidence points to a phenomenon of lipid–protein interaction, most likely of non-specific character. Indeed, an increasing body of experimental evidence suggests that lipid–protein interaction, beyond the involvement of an annular region of specific lipid molecules believed to be crucial for the functioning of certain channels (see, e.g., Lee, 2009; Schmidt et al., 2006), plays a crucial role also at a nonspecific level in the distribution as well as functional behavior of transmembrane proteins (Seeger et al., 2010; Alessandrini & Facci, 2011).

These kinds of experiments, therefore, provide support for the idea that the transmembrane proteins and lipids behave as a single molecular system endowed with given conformational and functional properties. Particularly, voltage gating and ion channel functional behavior is intrinsic to this binary system rather than to the ion channel per se, providing a concrete exemplification of the concept of system biology.

One could argue that it is hard to identify with a high degree of certainty the shape-changing bumps as Kat1 channels and that a positive identification of those molecules is needed to be safe in attributing the apparent voltage-responsive behavior to them. This is indeed an important point that unfortunately is not easy to tackle. One could, for instance, think of using specific antibodies to identify the channels. However, attempts in that direction have so far proved unsuccessful; indeed, it appears quite difficult to take advantage of immunolabeling when one has to deal with ion channels inserted in the membrane, as they are almost completely buried in the bilayer and epitopes that are regularly recognized in other conditions (e.g., Western blot) may be not available in more native situations.

Furthermore, the presence of large macromolecules, such as IgGs, could easily impair the conformational modifications of ion channels conjugated with them, making the experimental attempts described here completely vain.

Whereas inducing conformational modifications in voltage-sensitive ion channels by means of an external electric field can be considered, after all, not a big novelty, changing conformation and altering the affinity of antibodies for their corresponding antigens by a similar means is instead much more unexpected. That is what the next sections are about.

5.2 Direct electrical control of antibody conformation and affinity

A further, relevant framework where electrical control of biomolecular conformation and function can be important is that of biological molecules that per se are neither redox nor electric field responsive. With such a definition we refer to biomolecules whose functionality is usually not described in terms of an interaction with electric fields. Of course, this does not mean they cannot be affected, to a variable extent, by the action of such fields. Indeed, one has always to keep in mind that proteins are, generally speaking, zwitterionic macromolecules and that the charge state of their amino acid residues plays a major role in determining both their conformation and function (Branden & Tooze, 1999).

Our aim here is to show the potential of electrical control over biological entities and reactions and, for such a goal, a further illuminating example to devise strategies to implement control over the functionality of molecules such as antibodies that well represent biological constituents whose function is extremely important from both a physiological and a technological standpoint. Indeed, they mediate the immunological response and are widely used in many technological contexts for revealing and quantifying the presence of various antigens; furthermore, they represent the key molecular ingredient of any immunosensor, irrespective of the transduction mechanism chosen, imparting biochemical specificity to transduction elements (Campbell, 1991).

The implementation of direct electrical control of the functionality of antibodies can pave the way to a number of applications, which may include smart biosurfaces such as bio-adhesive surfaces "on demand", biomolecular-based memories, switchable immunosensing areas, etc. However, its implementation poses apparently unsolvable problems and requires unconventional approaches and solutions. That is why, in the literature, few attempts have been reported so far with this aim; definitely, the most relevant is that of U. Sivan's group, which reported the implementation of electrical control over an antibody–antigen reaction (Brod et al., 2008). In that work, the antibody–antigen interaction was monitored with the electrochemical surface plasmon resonance (SPR) technique. The antigen was immobilized on the working electrode while the antibody was injected in solution. After binding, application of a bias more negative than −0.5 V versus the Ag/AgCl reference electrode caused rapid detachment of the antibody molecules from the antigens. Removal of the applied voltage restored the antigen ability to bind the antibody molecules. The mechanism underlying the reported phenomenon was traced back to deprotonation of positively charged amino acids, particularly lysine, by hydroxyl ions generated at the electrode/solution interface. In spite of its interest, however, the reported approach relies on surface immobilization of antigens and on subsequent antibody capture/stripping. As such, it appears of limited technological interest for immunosensor application, where one seeks typically the presence and quantification of antigens in solution. Nevertheless, this work sets the scene for further development of the general concept of making antibody surfaces able to respond to an applied electric field, modulating thus their affinity for the corresponding antigens.

To move further on towards the aforementioned goal, however, a number of aspects deserve critical attention.

The first one is, of course, the choice of the molecules. There exist various types of antibodies that can have a monomeric (e.g., IgD, IgE, IgG), dimeric (e.g., IgA), or pentameric (e.g., IgM) structure. Surface immobilization strategies have been developed in the literature for all of these classes of antibodies (Caiazzo et al., 2009). A general enough possibility appears to be that of focusing on IgG-type antibodies, as the monomeric basic constituents of various antibodies and in view of their wide technological use.

In order to implement electrical driving of antibody–antigen reaction at surfaces, one has to rely on (i) proper surface biofunctionalization and (ii) direct electrochemistry. With these two basic ingredients, it is possible to drive electrically the conformation of surface immobilized antibodies, achieving, as a direct consequence, a modulation of their binding affinity for the corresponding specific antigens. To achieve such a goal, however, it is necessary to immobilize IgG-type molecules in a unique orientation in such a way that specific recognition sites are exposed to the solution, and to find the conditions to affect the IgG conformation with an electric field generated in solution by an electrically polarized metal substrate on which antibodies have to be immobilized.

5.2.1 The supramolecular construct

After the choice of the antibody type, a second, fundamental, aspect to focus on is the molecular immobilization strategy to follow in order to assemble IgGs on a surface. In this case, along with the usual requirements of a firm yet activity-preserving surface functionalization, one has to pay particular attention to antibody orientation. Indeed, this aspect has a major relevance in the specific case; indeed, proper antibody orientation allows for an optimal exposure of the antigen-binding regions on the Fab fragments to the solution containing antigens. Furthermore, a proper molecular orientation is also essential for an electric field to act effectively on the antibody conformation.

In order for an applied electric field to act effectively on surface immobilized molecules in solution, it is convenient to make a metal electrode which also plays the role of the substrate where antibodies are immobilized. This is the approach that we are going to follow to attempt this complex task.

If one chooses gold as the material for the electrode, one is quite naturally biased toward thinking of a thiol-based surface chemistry approach to achieve molecular immobilization. This is of course a plausible possibility but as such it does not provide an easy solution for imparting a specific orientation to the antibodies to be immobilized. Indeed, it is worth noting that a specific molecular orientation can be achieved only if one relies on reaction groups that can follow a single reaction pathway (i.e., reactions that take place between groups present in a single copy on both the molecule to be immobilized and the surface). If more than one copy of these reacting sites is present on the molecular surface, then the final protein orientation becomes unpredictable or not unique. That is the reason for seeking specific conjugation reactions.

When one deals with IgGs, there exists a quite straightforward and effective way of binding them to a surface by means of proteins that have the functional activity of binding F_c fragments with high affinity. These proteins are, for example, bacterial protein A and protein G. Protein A from *Staphylococcus aureus* possesses five F_c binding sites and can be commercially found in recombinant, mutated form (expressed in *E. coli*) with some deletion for nonspecific adsorption sites and, most interestingly, with a sequence of six His residues attached to its termini. Especially this last feature can orient one's choice. Indeed, this very common tag, which is inserted in many recombinant proteins for purification purposes, can help solve the problem of achieving an oriented IgG immobilization.

It is well known that His-tag has a high affinity for reagents such as Ni-NTA, which finds typical application in the fabrication of affinity chromatography columns. Less well known is the fact that the tag binds readily onto metal Ni surfaces and also onto Au ones. Figure 5.5 shows quartz crystal microbalance (QCM) adsorption kinetics of protein A on different surfaces, namely a Au surface pretreated so that it targets primary amines (2-mercaptoethylamine + glutaric dialdehyde), a Ni surface and a bare Au surface. As is clearly visible, not only does the His-tagged protein A bind the gold surface very well, but it does it with a much higher saturation level (i.e., surface density) than the more conventional, nonspecific amine-targeting chemistry. These results suggest the use of mutant protein A on gold electrodes as an optimal solution for our immobilization issues. Indeed, besides providing a very high surface density, probably brought about by the

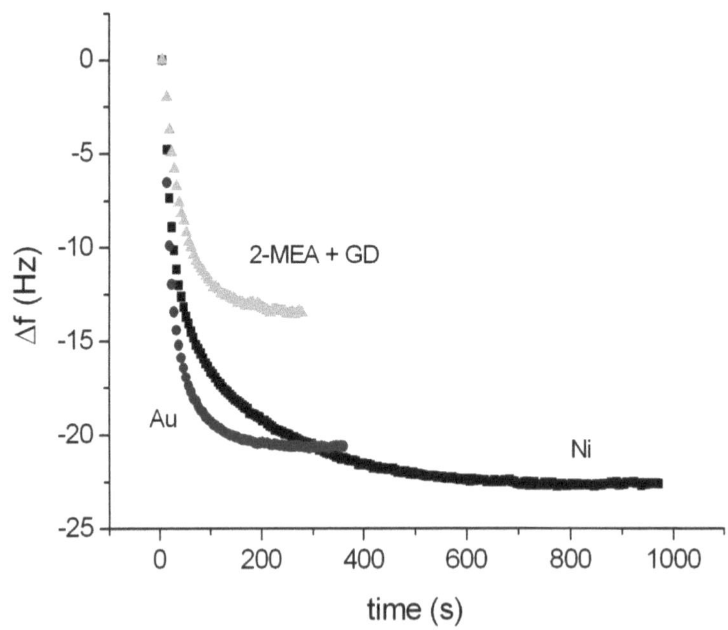

FIGURE 5.5

Comparison of adsorption kinetics of protein A on different substrates, namely functionalized Au (2-mercaptoethylamine + glutaric dialdehyde), bare Au and Ni.

higher diffusional degree of freedom that the molecule has on the surface with respect to covalent immobilization, exploiting His-tag for binding protein A to the surface has the remarkable advantage of defining a unique reaction pathway, since that involving the tag is the only effective way the protein can follow to bind to the gold surface. As a consequence, protein A becomes attached to the surface in a unique orientation.

Having settled this first part of the immobilization procedure, the rest is quite straightforward. Exposing the protein A-coated substrate to an IgG solution, antibodies bind readily to the adhesive adlayer. Due to the fact that protein A has, at its surface, an orientation determined by the position of the His-tail and that F_c adsorbs in five specific sites at the protein A surface, the resulting IgG layer appears to be endowed with a preferential orientation. We do not know exactly what that orientation is, but the results that we are going to report in what follows definitely support the idea that IgGs immobilized by the described approach possess a preferential orientation at the surface. Indeed, indirect evidence connected with the surface density of adsorbed IgGs is consistent with the formation of a uniaxially ordered protein layer. Figure 5.6 shows a comparison between the adsorption kinetics

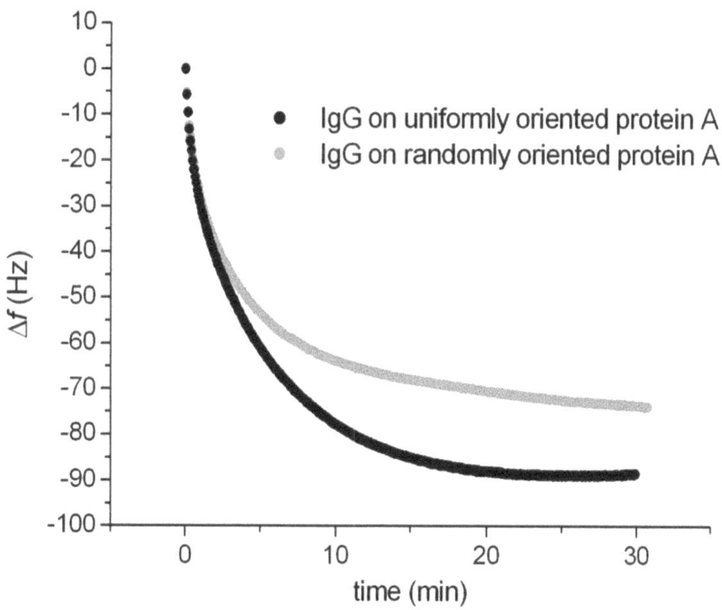

FIGURE 5.6

Comparison between adsorption kinetics of IgGs on a layer of His-tag oriented versus a randomly oriented layer of protein A.

of human IgGs bound to a QCM surface with two different treatments. The first one uses an adlayer of protein A immobilized exploiting surface amines (2-mercaptoethylamine + glutaric dialdehyde), whereas the second takes advantage of the His-tag approach.

The results reveal a 20% higher IgG coverage in the case of the oriented molecular construct. Interestingly, once the corresponding antigen has been exposed to the antibody layers, the oriented construct appears to capture 30% more antigens than the randomly oriented one. This difference points to an effect of steric hindrance the IgGs of the latter construct experience due to their random orientation on the surface; indeed, they are more prone to have some F_{ab} fragments hidden by neighboring molecules on the surface, at variance with the IgGs of the oriented construct that possess the aforementioned uniaxial order.

5.2.2 Implementing electric control over IgG conformation

An oriented IgG construct is just what one needs to proceed towards the development of a method for gaining technological control of protein conformation. Indeed, a possible solution to that problem lies through the exploitation of protein surface charges and their interaction with an external electric field. In the proposed set-up, an electric field can act on the molecular layer at the surface, being generated by the metal substrate supporting the film that acts as a working electrode of a three-electrode electrochemical cell and is connected to an external potentiostat.

For such a field to act effectively on the molecular layer, some key criteria have to be met. The first one is that the macromolecule at issue has some charged amino acid residues. This is easily achievable by inducing protonation of the primary amine moieties exposed to solution at the IgG surface. The typical IP of IgGs ranges between 6 and 9. Therefore, it is enough to buffer the working solution at pH values below 6 to ensure the presence of positive charges (essentially due to lysines) on the molecules. Another criterion is connected with molecular orientation, since only an overall (at least) uniaxial orientation can enable the layer to respond uniformly to any modulation of the electric field. Such a modulation can involve its intensity and its sign (direction) although it is always perpendicular to the plane of the substrate. Given a uniaxial IgG orientation (with the F_c fragment pointing towards the substrate), a modulation of field direction can have a push-pull effect on the IgGs (mediated essentially by their charged surface residues), probably involving the motion of F_{ab} fragments about their flexible hinges. This is, in a nutshell, the basic idea of the proposed approach. Figure 5.7

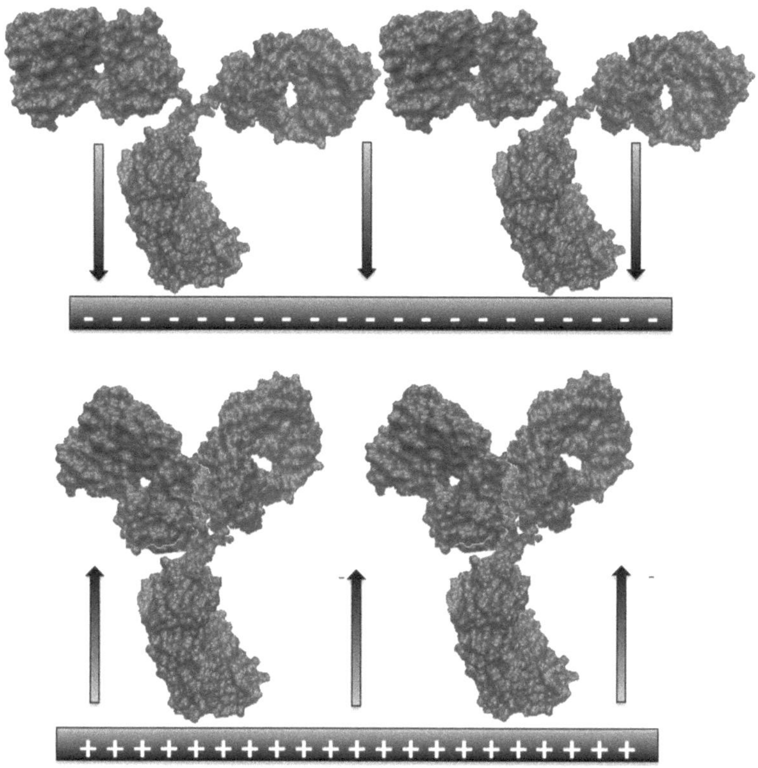

FIGURE 5.7

Schematic arrangement of the molecular orientation and the related effect of changing the sign of the substrate potential.

shows a simplified scheme of the system as affected by changing the direction of the field.

There is, of course, a further key criterion to meet, that is, to adjust the ionic strength of the operating solution in order to elicit a Debye length long enough to affect the total film thickness (see section 2.2). This is, clearly, a fundamental aspect of the overall enterprise, since the electric potential generated by the substrate has to be effective on the F_{ab} fragments located about 10 nm off the electrode surface. Ionic strength must be set, however, at a high enough value in order for the potentiostatic control to function and for the molecules involved to feel an environment not too different from the physiological one. Proper consideration of all these aspects suggests that 5 mM is the (equivalent) ionic strength to be used in this context, enabling a Debye length of about 10 nm (see section 2.2).

In order to test experimentally the effect on the supramolecular construct of a modulation of the potential value imposed at the working electrode, one can perform electrochemical QCM measurements as a function of the sign and intensity of the applied potential. The aim here is to seek the signature of a change in the conformation of the antibody layer as induced by electric potential modulation. Having immobilized a layer of molecules on one electrode of a QCM, we have measured its response to an electric potential changing in sign and intensity (100–200 mV). Figure 5.8 shows the corresponding data. Variations in both resonance frequency and FWHM (proportional to dissipation) displayed a similar, although opposite in direction, stepwise behavior upon changing electrode potential. These data suggest qualitatively that the layer is responsive to these changes. One can think of potential-induced conformational changes increasing the apparent viscoelastic load measured by QCM since positive charges, hence molecular domains, are attracted towards the electrode surface when it is negatively biased.

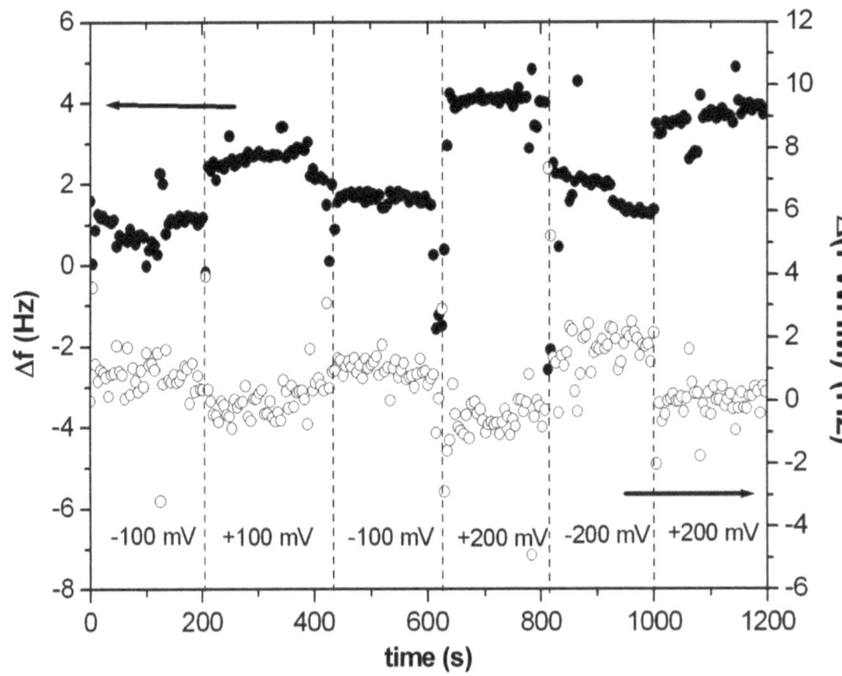

FIGURE 5.8

Electrochemical QCM measurements of an oriented antibody layer as a function of different substrate potential values.

These data just confirm that something meaningful happens at the molecular layer upon varying the potential value it senses. In order to deepen our understanding of the phenomena going on in the proposed set-up, another technique can be useful, namely, imaging the molecular layer by electrochemical scanning force microscopy. This is a special application of liquid scanning force microscopy where a potentiostat drives the potential value of the substrate while the microscope records the sample topography as in a conventional SFM experiment. The rationale of these measurements is to image a surface hosting the oriented molecular construct in a differential fashion, that is, at two different values of substrate potential, to seek topographical differences ascribable to IgG conformational variations induced by the applied potential. Of course, since here one deals with scanning force microscopy imaging, the substrate has to be prepared with special care as to its roughness, in order to elicit the imaging of small details. Therefore, we have used mica sheets and the template stripping method (Alessandrini et al., 2008) to grow 100-nm-thick Au films with atomically flat terraces. Then, the orienting treatment described in the previous section has been applied and imaging in tapping mode at different values of substrate potential has been performed on selected surface areas.

It is indeed hard to detect tiny height changes (of the order of 1 nm) performing repeated surface scans of the same area. Nonetheless, an analysis of the cross-section profiles on the acquired pairs of images clearly reveals differences in the height of the molecular adsorbates that are compatible with the postulated effect of the different potential values on the molecular conformation of IgGs. Figure 5.9 shows two typical ECSFM images along with the corresponding cross-section analysis and the histogram of the height differences of the two images.

These data are indeed compatible with a potential-induced height variation of the IgGs at the surface as due to the pushing effect of the applied electric field.

Although images represent in general convincing evidence of a phenomenon, it is difficult to retrieve all the needed information from a single type of measurement when one deals with such tiny, molecular effects. Therefore, it is usually very useful to collect data from different techniques to try to obtain concurrent evidence of the occurrence of a phenomenon. In this context, direct electrochemistry measurements, such as electrochemical impedance spectroscopy (EIS), can provide a valuable input. Indeed, it is possible to perform measurements on the construct at issue keeping the working electrode constituting our substrate at different base potentials, namely −100 and +100 mV, and seeking a difference in the response. Any difference in the response of such a system could be ascribed to conformational

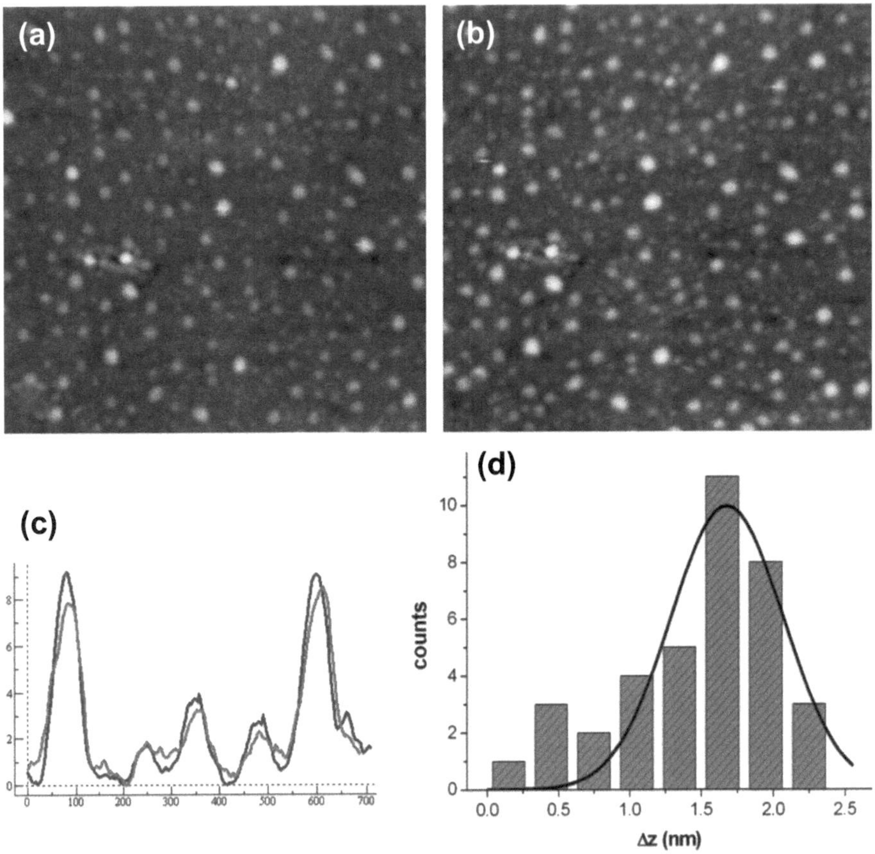

FIGURE 5.9

Two ECSFM images acquired at (a) −100 mV and (b) +100 mV. (c) A typical height profile. (d) Histogram of the height differences between the two images.

variations in the layer. It is interesting to note that, whereas the response of the layer is very different at the two base potentials investigated when the system is kept at pH 5.45, it does not change at all at pH 9. The results can be interpreted with an equivalent circuit that points to a variation in the layer capacitance upon changing potential, corroborating the model of conformational changes in the IgG layer, possibly due to motion of the F_{ab} fragments (even if the possibility of change in overall molecular orientation cannot be completely excluded from these data). Most importantly, the results obtained at pH 9, when IgGs are not charged, confirm the overall mechanism underlying the system as connected to the molecular charge state and consequent response to an applied electric field.

5.2.3 Implementing electric control over IgG functionality

Whereas structural information on the antibody layer is essential to try to understand the exact mechanism that is operating when we vary the potential at the working electrode, we should keep in mind that our final goal is to affect biological reactions, that is, in this specific case, to modulate antibody–antigen affinity. That is why it is essential to provide further evidence of the functionality of the antibody layer as connected to the potential-induced modulation of the IgG conformation.

Towards that further aim, electrochemical quartz crystal microbalance measurements can again turn out to be very useful. Indeed, one can perform adsorption kinetics measurements of specific antigens keeping the antibody layer at different base potentials to seek possible differences in the layer functional properties. Figure 5.10 shows data corresponding to such measurements. One notes that, at variance with a general uniformity in the kinetics aspects of the measurements performed at different base potentials, the saturation level, i.e., the maximum antigen amount captured by the layer, varies markedly between the two situations. This evidence points to modulation of the functional activity of the antibody layer as a function of the substrate potential, that is, as modulated by the different electrically induced IgG

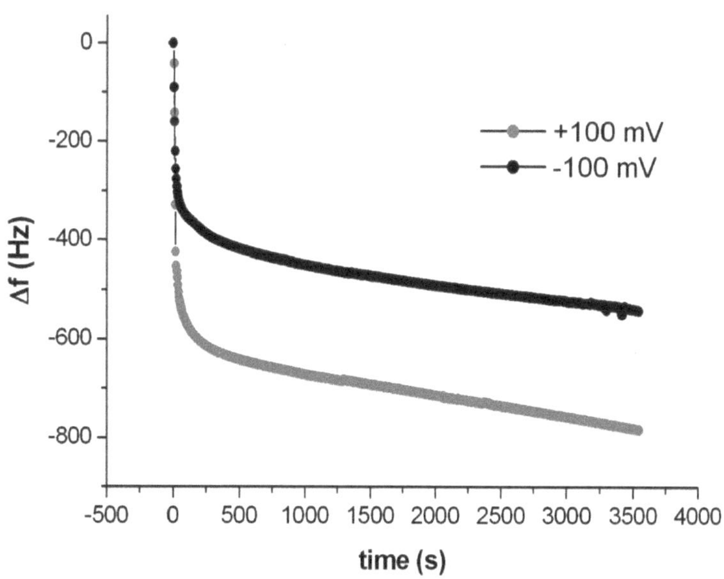

FIGURE 5.10

Comparison of antigen adsorption kinetics of two oriented protein A layers kept at different substrate potentials.

conformations. As in a previous case, changing solution pH further confirms the electrical origin of the phenomenon. Furthermore, molecular orientation is also involved in the picture, enabling modulation of the layer functionality. Indeed, the results of antigen binding kinetics measurements performed at two different base potentials on a randomly oriented IgG layer do not show any functional modulation. These data corroborate the overall nature of the picture that now starts to appear quite clearly.

Before drawing final conclusions, it is once more useful to use imaging to try to get a direct insight into the phenomenon of the electrical modulation of the functional activity of oriented antibody layers. It is indeed possible also to rely on fluorescence to further confirm the phenomenon. Briefly, one possibility is to use fluorescently labeled antigens, e.g., FITC-labeled insulin, and a substrate over which two independent Au electrodes have been defined by optical lithography. After having assembled the supramolecular construct on both the electrodes, with the help of a bipotentiostat, it is enough to incubate the antibody-coated electrodes keeping one at a positive (say, +100 mV) and the other at a negative (e.g., −100 mV) potential. After that, one images both electrodes and extracts the integrated fluorescence signal from each electrode. Figure 5.11 shows such images along with the normalized, integrated fluorescence values over the pairs of electrodes. It is clear that the electrode kept at a positive potential has captured about 40% more antigens than that kept at a negative potential value. Whereas the exact amount of antigens captured by a layer based at a positive potential with respect to that kept at a negative potential depends, among the other parameters, on the size of the antigens used, the trend is absolutely general. It is likely that the difference in the capturing ability among layers kept a different potentials can be optimized by finely tuning the antibody surface density (achievable, for example, by co-adsorbing complete IgG molecules and only F_c fragments). At any rate, these results support the concept that antibodies, properly oriented onto an electrode surface and charged thanks to the pH value of the solution, respond to the action of an external electric field modulating their conformation (most likely varying the orientation of their F_{ab} fragments) and as a consequence their affinity for antigens in solution (thanks to the different effects that steric hindrance can have on the attainable conformations).

The reported results extend the control of biological reactions by direct electrochemistry well beyond the framework of the redox reactions, paving the way to further generalization of this approach involving, for example, the modulation of the activity of surface-bound enzymes. Some examples of this kind have started to appear in the literature, albeit the level of molecular control implemented in them is still quite limited.

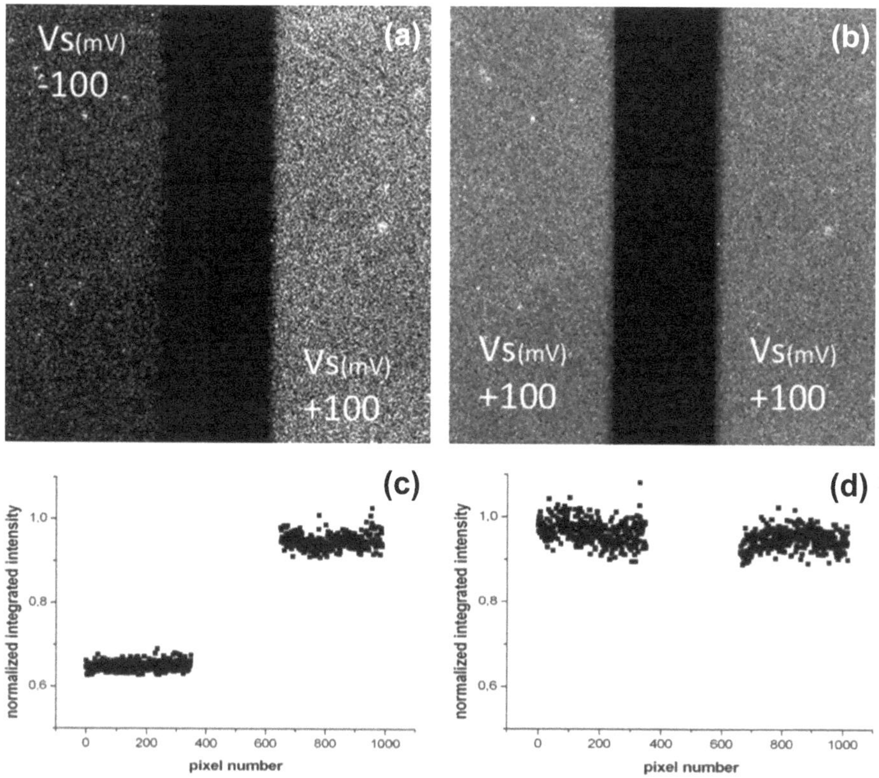

FIGURE 5.11

Florescence imaging of a pair of electrodes coated with oriented anti-insulin antibodies kept at (a) −100 mV (left) and +100 mV (right) and while incubated with FITC-labeled insulin; and (b) both at +100 mV. Panels (c) and (d) report the corresponding integrated intensities over the electrodes, separately.

5.3 Towards direct electrical modulation of enzyme activity

An electric enzyme regulator has recently been proposed (Chao et al., 2012) that takes advantage of a covalent, primary amine-targeting surface chemistry to immobilize the enzyme phenol sulfotransferase onto SiO_2 surfaces. In the reactor, the enzyme layer appears to be sandwiched between Pt and Au electrodes. Applying an electric stress to the enzyme layer by polarizing the structure with electric potential values of different sign and intensity results in a modulation of the enzyme activity. Particularly, negative potentials (at the enzyme layer surface) tend to decrease enzyme activity due to its electro-compression towards the substrate, whereas the contrary happens for positive potentials. This effect is traced back by the authors to the charged state of the molecule, which possesses an IP of 5.20–5.66, whereas the working solution is buffered at pH 7. The described situation recalls that described in the case of IgGs, albeit here with a much less defined molecular orientation.

Similarly, the activity of other enzymes (e.g. phospholipase A_2) has also been shown to depend upon the action of constant or varying electromagnetic fields to which they are exposed (Maggio, 1999). The difference in the kind of enzymes and in the context in which they operate suggests that the possibility of their electrical modulation is quite large and that it is likely that further examples will appear soon in the literature.

To sum up, the examples reported in this chapter foster the concept of technologically driven smart biosurfaces that can find application in a number of different (nano)technological contexts. Starting from the next chapter, we will enter the charming, largely unexplored, territory of the control of biological reactions in living beings. Then, the concepts developed here will act as a useful starting point towards much more ambitious goals.

5.4 References

Alessandrini, A., & Facci, P. (2005). AFM: a versatile tool in biophysics. *Meas. Sci. Technol., 16*, R65–R92.

Alessandrini, A., & Facci, P. (2011). Unravelling lipid/protein interaction in model bilayers by Atomic Force Microscopy. *J. Mol. Rec., 24*, 387–396.

Alessandrini, A., Bortolotti, C. A., Bertoni, G., Vezzosi, A., & Facci, P. (2008). Ultra-flat nickel substrates for scanning probe microscopy of His-tagged proteins. *J. Phys. Chem. C, 112*, 3747–3750.

Bashford, D. (2004). Macroscopic electrostatic models for protonation states in proteins. *Front. Biosci., 9*, 1082–1099.

Branden, C., & Tooze, J. (1999). *Introduction to protein structure* (2nd ed.). New York: Garland.

Brod, E., Nimri, S., Turner, B., & Sivan, U. (2008). Electrical control over antibody–antigen binding. *Sens. Actuators B, 128*, 560–565.

Caiazzo, M., Alessandrini, A., & Facci, P. (2009). A highly selective, bio-functional surface for molecule/cell sorting. *ACS Appl. Mater. Interfaces, 1*, 514–518.

Campbell, A. M. (1991). *Monoclonal antibody and immunosensor technology.* Amsterdam: Elsevier Science.

Chao, C.-H., Li, K.-L., Wu, C.-S., Lee, C.-C., Chiang, H.-P., Yang, Y.-S., Pan, T. M., & Ko, F. H. (2012). Surface effect of assembling enzyme and modulation of surface enzyme activity with electric potential stress. *Int. J. Electrochem. Sci., 7*, 5100–5114.

Dascal, N. (1987). The use of *Xenopus* oocytes for the study of ion channels. *CRC Crit. Rev. Biochem., 22*, 317–387.

Facci, P., Alliata, D., Andolfi, L., Schnyder, B., & Kötz, R. (2002). Formation and characterization of protein monolayers on oxygen-exposing surfaces by multiple-step self-chemisorption. *Surf. Sci., 504*, 282–292.

Han, W., Lindsay, S. M., & Jing, T. (1996). A magnetically driven oscillating probe microscope for operation in liquids. *Appl. Phys. Lett., 69*, 4111–4113.

Hille, B. (2001). *Ion channels of excitable membranes* (3rd ed.). Sunderland, MA: Sinauer Associates.

Hoshi, T. (1995). Regulation of voltage dependence of the KAT1 channel by intracellular factors. *J. Gen. Physiol., 105*, 309–328.

Jackson, M. B. (2006). *Molecular and cellular biophysics.* New York: Cambridge University Press.

Jiang, Y., Lee, A., Chen, J., Ruta, V., Cadene, M., Chait, B. T., & MacKinnon, R. (2003). X-ray structure of a voltage-dependent K1 channel. *Nature, 423*, 33–41.

Latorre, R., Olcese, R., Basso, C., Gonzales, C., Muñoz, F., Costelli, D., & Alvarez, O. (2003). Molecular coupling between voltage sensor and pore opening in the Arabidopsis inward rectifier K+ channel KAT1. *J. Gen. Physiol., 122*, 459–469.

Lee, A. G. (2009). The effects of lipids on channel function. *J. Biol., 8*, 86.

Maggio, B. (1999). Modulation of phospholipase A2 by electrostatic fields and dipole potential of glycosphingolipids in monolayers. *J. Lipid Res., 40*, 930–939.

Picco, C., Bregante, M., Naso, A., Gavazzo, P., Costa, A., Formentin, E., Downey, P., Lo Schiavo, F., & Gambale, F. (2004). Histidines are responsible for zinc potentiation of the current in KDC1 carrot channels. *Biophys J., 86*, 224–234.

Schachtman, D. P., Schroeder, J. I., Lucas, W. J., Anderson, J. A., & Gaber, R. F. (1992). Expression of an inward-rectifying potassium channel by the Arabidopsis KAT1 cDNA. *Science, 258*, 1654–1658.

Schmidt, D., Jiang, Q.-X., & MacKinnon, R. (2006). Phospholipids and the origin of cationic gating charges in voltage sensors. *Nature, 444*, 775–779.

Seeger, H. M., Aldrovandi, L., Alessandrini, A., & Facci, P. (2010). KcsA functionality is affected by the physical state of the hosting lipid bilayer. *Biophys. J., 99*, 3675–3683.

Suchyna, T. M., Markin, V. S., & Sachs, F. (2009). Biophysics and structure of the patch and the gigaseal. *Biophys. J., 97*, 738–747.

CHAPTER

Redox Control of Gene Expression Level

6

CHAPTER OUTLINE

6.0 Regulation of gene expression level

Among the most relevant biological reactions that one can consider as potential targets for direct electrical/electrochemical control, we plan now to focus on those complex phenomena that rule the level of expression of certain genes. The problem of understanding the biochemical pathways and the mechanisms by which even the simplest organisms implement an effective control over this crucial aspect of their functioning is an extremely intricate one and does not apparently possess a general solution. Furthermore, due to the inherent level of complexity, which often involves myriads of steps and molecular effectors in interaction with a changing environment, a robust, mechanistic picture that describes the interacting molecular events at the basis of the regulation is available only for a very limited set of rather simple cases. One has, indeed, to consider that due to the complex essence of these phenomena, it is extremely likely that the possible pathways yielding a given result are very numerous and even redundant, hence not amenable to being described by a "linear" sequence of independent steps.

6.1 Gene regulation in bacteriophages

Among the best-known cases, an emblematic exemplification is represented by the so-called regulator proteins in prokaryotes. In both bacteria and

Biomolecular Electronics. http://dx.doi.org/10.1016/B978-1-4557-3142-8.00006-6

bacteriophages, there exist regulator proteins having the function of activators and repressors. Repressors bind DNA at a structural gene promoter, preventing RNA-polymerase access to it, and thus preventing transcription. Activators, instead, bind next to a promoter, assisting polymerase in binding to it and resulting in enhanced gene transcription.

Some of the most-studied prokaryotic regulators are those of bacteriophage lambda, and related phages. These kinds of phages produce two regulator proteins called originally repressor and Cro. (These names are, indeed, misleading since it turned out lambda Cro is a repressor protein and lambda repressor can act as both repressor and activator. Nonetheless, they have been kept for historical reasons.) The two proteins operate a switch between two states of lambda phage replication.

We will consider now the molecular mechanism of action that these regulator proteins deploy to achieve gene control. Certain strains of *E. coli*, following irradiation with moderate doses of UV light, stop their normal growth and begin producing bacteriophages that eventually lead to bacterium lysis. These strains are called lysogenic and have the dormant DNA of the phage integrated in their own genome. In normal conditions the phage DNA is replicated along with the rest of the bacterial chromosome, but phase genes are not expressed. Irradiation with UV light switches the phage genes, which then produce new phages, leading eventually to bacterium death by lysis.

At a molecular level the situation is now quite well understood (Branden & Tooze, 1999). A rather small region of the phage genome contains all the components of the on-off switch.

Particularly, the region contains two structural genes that encode for the two regulator proteins Cro and repressor that regulate the switch, and the operator region (OR) on which they both act. The two ends of the OR are occupied by the promoters of the two genes, which are transcribed in opposite directions. When RNA polymerase is bound to the left-hand promoter, repressor is switched on, Cro and lytic genes become repressed, and the bacterium survives as a lysogenic strain. When, instead, RNA-polymerase is bound to the right-hand promoter, Cro is switched on along with the lytic genes that lie on its right, and cell lysis takes place. The lysis-lysogenic balance depends on which of the two promoters is able to bind polymerase and, in turn, that depends upon the binding of repressor and Cro proteins to three binding sites, OR1, OR2 and OR3, in OR. These binding sites are located in the middle of OR in such a way that OR1 and OR2 overlap the promoter of the Cro gene, and OR2 and OR3 that of the repressor gene. Both Cro and repressor bind to all three sites as dimers, but Cro has the highest affinity for OR3 and, when bound, it prevents RNA polymerase binding to the promoter of the repressor gene. On the other hand, repressor binds with highest affinity to OR1 and, when bound, it prevents polymerase accessing the Cro promoter.

Therefore, repressor establishes the lysogenic state binding to OR1 and switching off the lytic genes located at the right of Cro. Furthermore, it stabilizes this state by binding to OR2 and helping polymerase to bind, producing, thus, more repressor proteins. In other words, repressor acts in both repressing Cro synthesis and enhancing its own synthesis.

In contrast, Cro acts merely as a repressor, binding to OR3 with high affinity and preventing polymerase accessing the left-hand promoter and the consequent repressor synthesis.

Therefore, for lysogenic bacteria to be induced to producte phages, repressor must be released from OR1 in order for Cro to be synthesized and bind to OR3. This is exactly the effect of UV irradiation; indeed it activates the bacterial protease Rec A that cleaves the repressor dimers in two, preventing them from being stable dimers. Protease is activated when polymers of Rec A bind ssDNA as a consequence of DNA damage by UV photons. Since cleaved dimers have much less affinity for OR1 and OR2 than the intact counterparts, this frees the Cro promoter site for RNA-polymerase binding and transcription of the Cro gene and genes for viral replication. The newly synthesized Cro molecules bind thus to OR3, blocking the access of RNA-polymerase to the repressor promoter and preventing further repressor synthesis. From now on the switch has been flipped and the fate of the bacterium becomes lytic, due to diverging production of viral particles.

The situation in general, and in higher-level organisms in particular, is much more complex and ill-defined and as such it does not allow any simple description. Among the cases that apparently allow for a mechanistic understanding of the main steps of the regulation, one can find examples that particularly strike our interest. This is the case of the gene expression regulation mechanisms of some bacteria such as photosynthetic ones of the genus *Rhodobacter* as well as others such as *Escherichia coli*. Indeed, the identified mechanisms possess the interesting characteristic of being dependent upon the intracellular redox state. This fact, which we will try to dissect in its main aspects in the following sections of this chapter, is of course extremely intriguing for the aims of this book, since it opens up the charming possibility of starting to think of direct electrochemical control of rather complex biological processes even in living beings. Furthermore, as we will see after having reviewed the case of bacteria, also vegetal organisms, and green plants in particular, appear to be subjected to redox state regulation as far as the expression of the genes encoded in the genome of chloroplasts and mitochondria is concerned. We will focus on this topic in the second part of the present chapter.

6.2 Redox regulation of gene expression level: the case of *Rhodobacter*

The members of the genus *Rhodobacter* are facultative photosynthetic bacteria that synthesize photosynthetic complexes (see section 3.7.2) only if the oxygen tension in their environment is below a certain threshold. Oxygen affects expression of photosynthetic genes at transcriptional and posttranscriptional levels. The genes encoding for pigment-binding proteins are organized in two polycistronic operons: the *puf* operon, which encodes the pigment-binding proteins of the light-harvesting LHI complex and of the reaction center, and the *puc* operon, which encodes proteins of the LHII complex.

Many proteins are involved in oxygen-dependent transcription of these operons (Gregor & Klug, 1999, 2002) but a key role in transcription regulation of numerous genes of the aforementioned operons has been attributed to thioredoxins, small ($\approx$ 12 kDa) ubiquitous proteins containing an extremely reactive dithiol-disulfide in their active center, which are implicated in many cellular processes. However, till recently, their specific role in regulating *puf* and *puc* expression remained elusive.

Advances in the elucidation of the molecular interactions behind the regulating role of thioredoxins stemmed from the observations that *Rodobacter capsulatus*, like *E. coli*, possesses genes for thioredoxin 1 (trxA) and 2 (trxC), whereas *Rodobacter sphaeroides* lacks the trxC gene. Furthermore, a *R. sphaeroides* mutant strain expressing lower amounts of TrxA than the wild-type counterpart shows lower *puf* and *puc* transcription levels following a transition from high to low oxygen tension (Pasternak et al., 1999). A *R. capsulatus* strain lacking the trxC gene shows high transcription levels of *puf* and *puc* genes under the same conditions if compared to its parental wild-type strain. In order to shed light on the signaling pathway leading from thioredoxin to photosynthesis gene transcription, an approach called the yeast two-hybrid system (Li et al., 2003) has been used to seek proteins that interact with thioredoxin and are likely to be involved in oxygen-dependent regulation of photosynthesis genes. Such an approach has revealed that thioredoxin interacts with a number of proteins, among which is the DNA gyrase B subunit (GyrB). The bacterial type IIA enzyme DNA gyrase is the only topoisomerase capable of introducing negative supercoils into DNA. From a structural standpoint, gyrase acts as an A_2B_2 tetramer, encoded by the gyrA and gyrB genes, respectively (Reece & Maxwell, 1991). An earlier study revealed that DNA gyrase activity affects the expression of *puf* and *puc* genes in *R. capsulatus* (Zhu & Hearst, 1988).

As just mentioned, gyrase catalyzes the introduction of negative supercoils into DNA (Reece & Maxwell, 1991). The protein binds to DNA in the form of a tetramer (two A and two B subunits) and wraps DNA into a positive supercoil. One region of the DNA duplex is passed through another by means of DNA breakage and religation, as is typical for a topoisomerase. The binding of ATP drives the supercoiling reaction, with ATP hydrolysis serving to reset the enzyme for a second round of catalysis. In the absence of ATP, gyrase also catalyzes the relaxation of negatively supercoiled DNA in an ATP-independent reaction. Since the ratio of ATP to ADP determines the final level of supercoiling (Westerhoff et al., 1988; Drlica, 1992), this makes gyrase activity sensitive to modifications in intracellular energetics and, as a consequence, to the extracellular environment. Gyrase is also known to facilitate the movement of transcription complexes through DNA by introducing negative supercoils ahead of it and can influence transcription by DNA bending and folding (Malik et. al., 1996).

As a result, gyrase can affect transcription rates by its supercoiling or relaxation activities, since the local supercoiling of DNA can, in its turn, influence promoter activities (Whitehall et al., 1993; Karem & Foster, 1993; Leclerc et al., 1998). It is fair to say that the influence of DNA supercoiling on gene expression can differ significantly for individual genes. Indeed, as already reported in the literature, increased supercoiling can stimulate gene expression, but can also decrease the gene expression level (see section 6.3).

Extensive investigations (Li et al., 2003), involving the study of gyrase supercoiling activities in vitro with cell extracts, have come up with an intriguing model that is the basis of our proposed approach to the technological (electrical) control of gene expression level.

The thioredoxin effect on gyrase functional activity appears to depend upon its redox state. According to this model, a variation in oxygen tension influences the redox state of thioredoxins, which, in turn, alters gyrase activity. Since it was found that oxidized TrxC inhibits gyrase activity, whereas reduced TrxA stimulates it, both thioredoxins mediate a signal leading to higher supercoiling activity at reduced redox potential. The opposite effects of TrxA and TrxC on the supercoiling activity of gyrase correlate well with the opposite effect of trxA and trxC mutations on the expression of the photosynthesis genes and bacteriochlorophyll accumulation after oxygen tension reduction in *Rhodobacter* (Pasternak et al., 1999; Li et al., 2003). Figure 6.1 summarizes this model in the case of *Rhodobacter*.

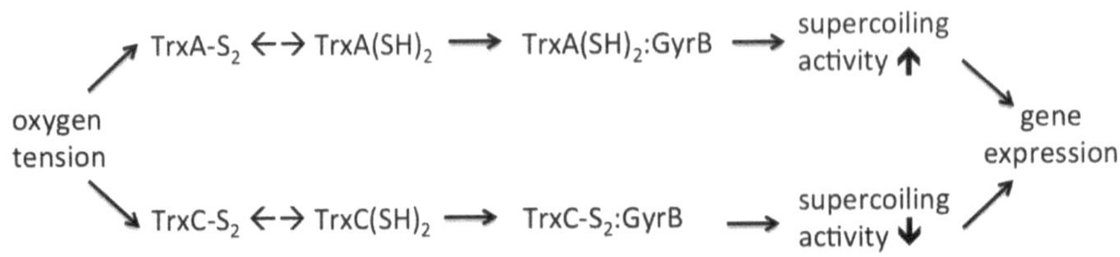

FIGURE 6.1

Schematic model for signal transduction from thioredoxin to gene expression in *Rhodobacter*. The redox switch of thioredoxins, induced by a change in oxygen tension, alters the supercoiling activity of gyrase, which further affects gene expression.

The case of *Rhodobacter* represents, for the aims of the present book, a paradigmatic example of a complex biological function, such as the modulation of gene expression level, naturally subjected to redox regulation. As such, it possesses a necessary feature that makes it a possible candidate to be subjected to a direct electrochemical drive. However, before entering into the details required for such an ambitious implementation, several other aspects have to be clarified that will be the subject of this and the next chapters.

6.3 Redox regulation of gene expression level: the case of *Escherichia coli*

Escherichia coli is perhaps the most studied and best known among the living organisms. It is currently used as a molecular factory for the expression of many different proteins (see, e.g., Lee, 2009), and in fact represents a sort of gold standard for biological research. Therefore, the fact that this bacterium also possesses redox-regulated genes represents a relevant peculiarity and entitled us to consider this type of gene expression level modulation a quite general mechanism used by bacteria in several different cases.

In a fashion very similar to the case of the aforementioned *Rhodobacter*, different studies have indeed shown the role of DNA supercoiling in regulating the level of expression of diverse genes. As we have already recalled in section 6.1, the influence of DNA supercoiling on gene expression can differ significantly for individual genes. Also in the case of *E. coli*, higher supercoiling can favor gene expression, and this is, for instance, the case for topoisomerase A (Tseh-Dinh & Beran, 1988) and the *fis* gene (Schneider et al., 2000). However, it can also have the effect of decreasing the gene expression level, as in the case of gyrase (Franco & Drlica, 1989).

In all these and other cases, the role of supercoiling as determined by the action of gyrases is affected by the redox state of thioredoxins. In the case of *E. coli*, the situation is, however, more complex due to the presence, besides thioredoxins (*trxA* and *trxC*), of three genes encoding for glutaredoxins (*grxA*, *grxB* and *grxC*). From a study in which *trxA*- and *trxC*-deleted *E. coli* were shown to be viable, it appeared that the glutaredoxin system can replace, from a functional point of view, the thioredoxin system. In light of this characteristic of *E. coli* and the redox-regulated modulation of photosynthetic pigment synthesis in the case of *Rhodobacter*, the latter appears as a more suitable candidate for designing a direct electrochemical approach to gene expression level control, as we will see in Chapter 7. Indeed, monitoring and quantifying the presence of bacteriochlorophylls is much easier and more convenient than any other test to assess the presence and the amount of a given gene, since it requires simple and direct spectrophotometric assays.

6.4 Redox control of gene expression in subcellular organelles

Chloroplasts and mitochondria are energy-converting organelles present in the cytoplasm of eukaryotic cells. Chloroplasts in plant cells perform photosynthesis, i.e., the capture and conversion of the energy of sunlight. Mitochondria in both plant and animal cells perform respiration, i.e., the release of stored energy to perform work. In addition to these key aspects, chloroplasts and mitochondria each contain specialized and discrete genetic systems. These genetic systems enable chloroplasts and mitochondria to synthesize some of their own proteins.

Both the genetic and energy-converting systems of chloroplasts and mitochondria derive, with little modification, from those of the free-living microorganisms that these organelles once were. The existence of these cytoplasmic genomes is consistent with, and provides evidence for, the endosymbiont hypothesis. According to this theory, mitochondria and chloroplasts (plastids in general), and possibly other organelles, represent formerly free-living bacteria that were taken inside another cell as an endosymbiont (see, e.g., Raven et al., 2010). Molecular and biochemical evidence suggests that mitochondria developed from protists, whereas chloroplasts came from cyanobacteria. Most genes for the proteins of chloroplasts and mitochondria are now, however, located on chromosomes in the nuclei of eukaryotic cells. There, they encode for protein precursors that are made in the cytosol for subsequent transfer into the organelles. This fact is not trivial and figuring out why some genes have moved to the nucleus, whereas others remain inside these organelles, is potentially very important for an understanding of the evolutionary development of eukaryotic cells. Apparently, there are no arguments that explain the reason for this apparently costly (more than 90 proteins have to be encoded for by nuclear DNA to make such a system work) subdivision of genes between the nuclear and cytoplasmic genomes, and thorough investigation of the organelles' genomes has provided no clues. However, a suggestive hypothesis has been put forward that tries to explain this peculiar fact relying on the particular redox state of the interior of the organelles. Such a hypothesis, called the CoRR hypothesis (acronym for Co-location for Redox Regulation), will be addressed in more detail in section 6.4.3. Here, it is enough to outline that the relevance of chloroplast and mitochondrion systems to the aims of this book is markedly dependent on the consequences of that hypothesis.

6.4.1 The mitochondrion in brief

The mitochondrion is a membrane-wrapped structure found in most eukaryotic cells (Henze & Martin, 2003). Mitochondrion diameter ranges from 0.5 to 1.0 μm. They are sometimes considered as "cellular power plants" because they generate most of the ATP supply for the cell. In addition to supplying cellular energy, mitochondria are involved in other tasks such as signaling, cellular differentiation, cell death, as well as the control of the cell cycle and cell growth (McBride et al., 2006). Mitochondria have been implicated in several human diseases, including mitochondrial disorders (Gardner & Boles, 2005) and cardiac dysfunction (Lesnefsky et al., 2001), and seem to play a role in aging processes.

Several characteristics make mitochondria unique. The number of mitochondria in a cell varies widely with organism and tissue type. Many cells have only a single mitochondrion, whereas others can contain several thousand of them (see, e.g., Alberts et al., 1994; Voet et al., 2006). The mitochondion is composed of compartments that carry out specialized functions. These compartments or regions include the outer membrane, the intermembrane space, the inner membrane, and the cristae and matrix. Mitochondrial proteins vary depending on the tissue and the species. In humans, 615 distinct types of proteins have been identified from cardiac mitochondria (Taylor et al., 2003), whereas in rats 940 proteins have been reported (Zhang et al., 2008a). The mitochondrial proteome is thought to be dynamically regulated (Zhang et al., 2008b). Although most of a cell's DNA is contained in the nucleus, the mitochondrion has its own independent genome. Further, its DNA shows substantial similarity to bacterial genomes (Andersson et al., 2003).

A mitochondrion contains outer and inner membranes composed of phospholipid bilayers and proteins (Alberts et al., 1994). The two membranes have different properties. Because of this double-membrane organization, there are five distinct parts to a mitochondrion. They are: (i) the outer mitochondrial membrane; (ii) the intermembrane space; (iii) the inner mitochondrial membrane; (iv) the cristae space (formed by deep corrugations of the inner membrane); and (v) the matrix (space within the inner membrane).

The outer mitochondrial membrane has a protein-to-phospholipid ratio similar to that of the eukaryotic plasma membrane (about 1:1 w:w). It contains large numbers of integral proteins called porins. These porins form channels that allow molecules less than 5000 daltons in molecular weight to freely diffuse from one side of the membrane to the other (Alberts et al., 1994). Larger proteins can enter the mitochondrion if a signaling sequence at their N-terminus binds to a large multi-subunit protein called translocase in the outer membrane, which then actively moves them across the membrane (Herrmann & Neupert, 2000). Disruption of the outer membrane permits proteins in the intermembrane space to leak into the cytosol, leading to certain cell death (Chipuk et al., 2006). The mitochondrial outer membrane can associate with the endoplasmic reticulum (ER), in a structure called MAM (mitochondria-associated ER-membrane), which is relevant in ER-mitochondria calcium signaling and involved in the transfer of lipids between the ER and mitochondria (Hayashi et al., 2009).

Because the outer membrane is freely permeable to small molecules, the concentration of small molecules such as ions and sugars in the intermembrane space is the same as in the cytosol (Alberts et al., 1994). However, this is not the case for large proteins, so the protein composition of this space is different from the protein composition of the cytosol. One protein that is localized to the intermembrane space in this way is cytochrome *c* (Chipuk et al., 2006).

The inner mitochondrial membrane contains proteins with five types of function (Alberts et al., 1994): (i) those that perform the redox reactions of oxidative phosphorylation; (ii) ATP synthase, which generates ATP in the matrix; (iii) specific transport proteins that regulate metabolite passage into and out of the matrix; (iv) protein import machinery; (v) mitochondria fusion and fission protein.

The inner membrane contains more than 151 different polypeptides, and has a very high protein-to-phospholipid ratio (more than 3:1 w:w, which corresponds to about 1 protein for 15 phospholipids). This membrane is home to around 1/5 of the total mitochondrial proteosome (Alberts et al., 1994). In addition, the inner membrane is rich in an unusual phospholipid, cardiolipin. This phospholipid was originally discovered in cow hearts in 1942, and is usually characteristic of mitochondrial and bacterial plasma membranes (McMillin & Dowhan, 2002). Cardiolipin contains four acyl chains rather than two, and may help to make the inner membrane impermeable (Alberts et al., 1994). Unlike the outer membrane, the inner membrane does not contain porins, and is highly impermeable to all molecules. Almost all ions and molecules require special membrane transporters to enter or exit the matrix. Proteins are transferred into the matrix via the translocase of the inner membrane (TIM) complex or via Oxa1 (Herrmann & Neupert, 2000). In addition, the action of the enzymes of the electron transport chain maintains a membrane potential across the inner membrane.

The inner mitochondrial membrane is organized into numerous cristae, which expand the surface area of the inner mitochondrial membrane, enhancing its ability to produce ATP. For typical liver mitochondria, the area of the inner membrane is about five times as large as the outer membrane. This ratio is variable, depending on the type of cell, and mitochondria from cells that have a greater demand for ATP, such as muscle cells, contain even more cristae. These folds are decorated with small round bodies known as F1 particles or oxysomes. These are not simple random folds but rather invaginations of the inner membrane, which can affect the overall chemiosmotic function (Mannella, 2006).

The matrix is the space enclosed by the inner membrane. It contains about 2/3 of the total protein in a mitochondrion (Alberts et al., 1994). The matrix is important in the production of ATP with the aid of the ATP synthase contained in the inner membrane. The matrix contains a highly concentrated mixture of hundreds of enzymes, special mitochondrial ribosomes, tRNA and several copies of the mitochondrial DNA genome.

Mitochondria have their own genetic material, and the machinery to manufacture their own RNAs and proteins. The human mitochondrial DNA sequence contains 16,569 base pairs encoding 37 genes in total: 22 tRNA, 2 rRNA and 13 peptide genes (Anderson et al., 1981). The 13 mitochondrial peptides in humans are integrated into the inner mitochondrial membrane, along with proteins encoded by genes that reside in the host cellular nucleus.

The most prominent roles of mitochondria are to store energy in chemical form in ATP (i.e., ADP phosphorylation) through respiration, and to regulate cellular metabolism (Voet et al., 2006). The central set of reactions involved in ATP production are collectively known as the citric acid cycle, or the

Krebs cycle. However, the mitochondrion has many other functions in addition to the production of ATP.

A dominant role for the mitochondria is the production of ATP, as reflected by the large number of proteins in the inner membrane for this task. This is done by oxidizing the major products of glucose, pyruvate and NADH, which are produced in the cytosol (Voet et al., 2006). This process of cellular respiration, also known as aerobic respiration, is dependent on the presence of oxygen. When oxygen is limited, the glycolytic products will be metabolized by anaerobic fermentation, a process that is independent of the mitochondria (Voet et al., 2006). The production of ATP from glucose has an approximately 13-times higher yield during aerobic respiration compared to fermentation (Rich, 2003). Recently it has been shown that plant mitochondria can produce a limited amount of ATP without oxygen by using the alternative substrate nitrite (Stoimenova et al., 2007).

6.4.2 The chloroplast in brief

Chloroplasts, present in plants and green algae, are the organelles deputed to photosynthesis; in many respects they are similar to mitochondria. Both chloroplasts and mitochondria have the function of producing metabolic energy, evolved by endosymbiosis, contain their own genetic systems and replicate by division. However, chloroplasts are more complex and larger than mitochondria, and perform several key tasks in addition to the generation of ATP. Particularly, chloroplasts are responsible for the photosynthetic formation of carbohydrates from CO_2. Furthermore, chloroplasts synthesize amino acids, fatty acids and the lipid components of their own membranes. In chloroplasts the reduction of nitrite (NO_2^-) to ammonia (NH_3) also takes place, an essential step in the incorporation of nitrogen into organic compounds. Moreover, chloroplasts are only one of several types of related organelles (plastids) that play a variety of roles in plant cells.

Plant chloroplasts are large organelles (5 to 10 μm in size) that, like mitochondria, are wrapped into a double membrane called the chloroplast envelope. In addition to the inner and outer membranes of the envelope, chloroplasts have a third internal membrane system, called the thylakoid membrane. The thylakoid membrane forms a network of flattened discs called thylakoids, which are frequently arranged in stacks called grana. Due to this three-membrane structure, the internal organization of chloroplasts is more complex than that of mitochondria. Particularly, the three membranes divide chloroplasts into three distinct compartments: (i) the intermembrane space between the two membranes of the chloroplast envelope; (ii) the stroma, lying inside the envelope but outside the thylakoid membrane; and (iii) the thylakoid lumen.

Despite this higher degree of complexity, the membranes of chloroplasts have remarkable functional similarities to those of mitochondria – as expected, since both organelles are involved in ATP generation. The outer membrane of the chloroplast, like that of mitochondria, contains porins that allow for permeation of small molecules. The inner membrane is impermeable to ions and metabolites, which can enter chloroplasts only by exploiting specific membrane transporters. These properties of the inner and outer membranes of the chloroplast envelope are similar to those of mitochondria: in both cases the inner membrane limits the passage of molecules between the cytosol and the interior of the organelle. The chloroplast stroma also plays a similar function to the mitochondrial matrix; indeed, it contains the chloroplast genetic system and a number of enzymes, including those responsible for the critical conversion of CO_2 to carbohydrates during photosynthesis.

The major difference between chloroplasts and mitochondria, in terms of both structure and function, is the thylakoid membrane. This membrane is of key importance in chloroplasts, where it plays the role of the inner mitochondrial membrane in electron transport and the chemiosmotic ATP generation. The inner membrane of the chloroplast (which is not folded into cristae, at variance with that of mithocondria) does not have a functional role in photosynthesis. Instead, the chloroplast electron transport system is located in the thylakoid membrane, and protons are pumped across this membrane from the stroma to the thylakoid lumen. The resulting electrochemical gradient drives ATP synthesis as protons cross back into the stroma. In terms of its role in the generation of metabolic energy, the thylakoid membrane of chloroplasts is thus equivalent to the inner membrane of mitochondria.

Like mitochondria, chloroplasts contain their own genetic system, believed to reflect their evolutionary origins from photosynthetic protists. The genomes of chloroplasts are similar to those of mitochondria because they consist of circular DNA molecules present in multiple copies in each organelle. However, chloroplast genomes are larger and more complex than those of mitochondria, ranging from 120 to 160 kb and containing approximately 120 genes.

The chloroplast genomes of several plants have been completely sequenced, leading to the identification of many genes contained in the organelle DNAs. These chloroplast genes encode both RNAs and proteins involved in gene expression, as well as a variety of proteins that function in photosynthesis. Both the ribosomal and transfer RNAs used for translation of chloroplast mRNAs are encoded by the organelle genome. At variance with the smaller number of tRNAs encoded by the mitochondrial genome, the chloroplast tRNAs are sufficient to translate all the mRNA codons according to the universal genetic code. In addition to these RNA components of the translation system, the chloroplast genome encodes for about 20 ribosomal proteins, which represent approximately a third of the proteins of chloroplast ribosomes. Some subunits of RNA polymerase are also encoded by chloroplasts, although additional RNA polymerase subunits and other factors needed for chloroplast gene expression are encoded in the nucleus.

The chloroplast genome also encodes for about 30 proteins that are involved in photosynthesis, including components of PSI and PSII, of the cytochrome *bf* complex, and of ATP synthase. In addition, one of the subunits of ribulose bisphosphate carboxylase (rubisco) is encoded by chloroplast DNA. Rubisco is the key enzyme that catalyzes the addition of CO_2 to ribulose-1,5-bisphosphate during the Calvin cycle.

Why do organelles encode a substantial number of genes in their own DNA, whereas others have been translated to the nuclear genome? A possible, interesting answer to this question is provided by the CoRR hypothesis.

6.4.3 The CoRR hypothesis: colocation for redox regulation

Why do mitochondria and chloroplasts have their own separate genetic systems, whereas other organelles that share the same cytoplasm, such as peroxisomes and lysosomes, do not? Such a question is definitely not trivial since, due to the energetic cost that such a gene arrangement brings about (more than 90 proteins are encoded in the nuclear DNA for this reason), it has to have profound reasons to be maintained by evolution.

A suggestive, possible answer to that question was proposed in 1993 by J.F. Allen with his CoRR hypothesis (Allen, 1993). Indeed, the hypothesis of "*colocation for redox regulation*" seeks to explain why chloroplasts and mitochondria retain DNA, and thus why some characters are inherited through the cytoplasm in the phenomenon of cytoplasmic, non-Mendelian, maternal inheritance.

In essence, the CoRR hypothesis states that chloroplasts and mitochondria contain those genes whose expression is required to be under the direct, regulatory control of the redox state of their gene products, or of electron carriers with which those gene products interact. Such genes comprise a primary subset of organellar genes. The need for redox control of each gene in the primary subset then confers an advantage upon location of that gene within the organelle. As a consequence, natural selection anchors some genes in organelles, while favoring location of others in the nuclear genome.

Chloroplast and mitochondrial genomes also contain genes for components of the chloroplast and mitochondrial genetic systems themselves. These genes comprise a secondary subset of organellar genes: genetic system genes. There is generally no need for redox control of expression of genetic system genes, although the fact that these are subject to redox control may, in some cases, allow amplification of redox signals acting upon genes in the primary subset (bioenergetic genes).

Retention of genes of the secondary subset (genetic system genes) is necessary for the operation of redox control of expression of genes in the primary subset. If all genes disappear from the primary subset, CoRR predicts that there is no function for genes in the secondary subset, and such organelles will then, eventually, lose their genomes completely. However, if even a single gene remains under redox control, then an organelle genetic system is required for the synthesis of its single gene product.

Among the experimental evidence that corroborates the CoRR hypothesis one can recall the fact that different products of protein synthesis in isolated chloroplasts and mitochondria are obtained in the presence of redox reagents with different redox potentials (Allen et al., 1995). In mitochondria, the effect results from a redox signal at the level of respiratory complex II.

Moreover, genes in chloroplasts are selected for transcription according to the redox state of the electron carrier plastoquinone (Pfannschmidt et al., 1997). These genes code for photosynthetic reaction centers and other components of the photosynthetic electron transport chain.

Furthermore, a modified bacterial sensor kinase couples transcription in chloroplasts to the plastoquinone redox state (Puthiyaveetil et al., 2008). This chloroplast sensor kinase is inherited from cyanobacteria and encoded in plant nuclear DNA. Chloroplast sensor kinase is part of a redox two-component regulatory system in eukaryotes (Puthiyaveetil & Allen, 2009).

The corroboration of the CoRR hypothesis along with the other evidence for redox-regulated gene expression in bacteria (see sections 6.2 and 6.3) sets the frame for the remaining part of this book. Indeed, the presence in different organisms and cases of redox-regulated genes, albeit characterized by varying levels of complexity, enables the idea that is the basis of the next chapters, i.e., the implementation of a direct electrochemical control of gene expression level.

Needless to say, the reported evidence is just one among several necessary conditions for attaining the abovementioned electrochemical control. In the next chapter we will see how we can try to fulfill other conditions in order to come to an acceptable level of feasibility for the proposed idea.

6.5 References

Alberts, B., Johnson, A., Lewis, J., Raff, M., Roberts, K., & Walter, P. (1994). *Molecular biology of the cell*. New York: Garland.

Allen, J. F. (1993). Control of gene expression by redox potential and the requirement for chloroplast and mitochondrial genomes. *J. Theor. Biol., 165*, 609–631.

Allen, C. A., Hakansson, G., & Allen, J. F. (1995). Redox conditions specify the proteins synthesized by isolated-chloroplasts and mitochondria. *Redox Report, 1*, 119–123.

Anderson, S., Bankier, A. T., Barrell, B. G., de-Bruijn, M. H. L., Coulson, A. R., et al. (1981). Sequence and organization of the human mitochondrial genome. *Nature, 290*, 427–465.

Andersson, S. G., Karlberg, O., Canbäck, B., & Kurland, C. G. (2003). On the origin of mitochondria: a genomics perspective. *Phil. Trans. R. Soc. B, 358*, 165–177.

Branden, C., & Tooze, J. (1999). *Introduction to protein structure* (2nd ed.). New York: Garland.

Chipuk, J. E., Bouchier-Hayes, L., & Green, D. R. (2006). Mitochondrial outer membrane permeabilization during apoptosis: the innocent bystander scenario. *Cell Death Differ., 13*, 1396–1402.

Drlica, K. (1992). Control of bacterial supercoiling. *Mol. Microbiol., 6*, 425–433.

Franco, R. J., & Drlica, K. (1989). Gyrase inhibitors can increase gyrA expression and DNA supercoiling. *J. Bacteriol., 171*, 6573–6579.

Gardner, A., & Boles, R. G. (2005). Is a "mitochondrial psychiatry" in the future? A review. *Curr. Psychiatry Rev., 1*, 255–271.

Gregor, J., & Klug, G. (1999). Regulation of bacterial photosynthesis genes by oxygen and light. *FEMS Microbiol. Lett., 179*, 1–9.

Gregor, J., & Klug, G. (2002). Oxygen-regulated expression of genes for pigment binding proteins in *Rhodobacter capsulatus*. *J. Mol. Microbiol. Biotechnol., 4*, 249–253.

Hayashi, T., Rizzuto, R., Hajnoczky, G., & Su, T. P. (2009). MAM: more than just a housekeeper. *Trends Cell Biol., 19*, 81–88.

Henze, K., & Martin, W. (2003). Evolutionary biology: essence of mitochondria. *Nature, 426*, 127–128.

Herrmann, J. M., & Neupert, W. (2000). Protein transport into mitochondria. *Curr. Opin. Microbiol., 3*, 210–214.

Karem, K., & Foster, J. W. (1993). The influence of DNA topology on the environmental regulation of a pH-regulated locus in *Salmonella typhimurium*. *Mol. Microbiol., 10*, 75–86.

Leclerc, G. J., Tartera, C., & Metcalf, E. S. (1998). Environmental regulation of *Salmonella typhi* invasion-defective mutants. *Infect. Immun., 66*, 682–691.

Lee, S. Y. (2009). *System biology and biotechnology of Escherichia coli*. Berlin: Springer Science.

Lesnefsky, E. J., Moghaddas, S., Tandler, B., Kerner, B., & Hoppel, C. L. (2001). Mitochondrial dysfunction in cardiac disease: ischemia–reperfusion, aging, and heart failure. *J. Mol. Cell. Cardiol., 33*, 1065–1089.

Li, K., Hartig, E., & Klug, G. (2003). Thioredoxin 2 is involved in oxidative stress defence and redox-dependent expression of photosynthesis genes in Rhodobacter capsulatus. *Microbiology, 149*, 419–430.

Malik, M., Bensaid, A., Rouviere-Yaniv, J., & Drlica, K. (1996). Histone-like protein HU and bacterial DNA topology: suppression of an HU deficiency by gyrase mutations. *J. Mol. Biol., 256*, 66–76.

Mannella, C. A. (2006). Structure and dynamics of the mitochondrial inner membrane cristae. *Biochim. Biophys. Acta, 1763*, 542–548.

McBride, H. M., Neuspiel, M., & Wasiak, S. (2006). Mitochondria: more than just a powerhouse. *Curr. Biol., 16*, R551–560.

McMillin, J. B., & Dowhan, W. (2002). Cardiolipin and apoptosis. *Biochim. Biophys. Acta, 1585*, 97–107.

Pasternak, C., Haberzettl, K., & Klug, G. (1999). Thioredoxin is involved in oxygen regulated formation of the photosynthetic apparatus of *Rhodobacter sphaeroides*. *J. Bacteriol., 181*, 100–106.

Pfannschmidt, T., Nilsson, A., & Allen, J. F. (1997). Photosynthetic control of chloroplast gene expression. *Nature, 397*, 625–628.

Puthiyaveetil, S., & Allen, J. F. (2009). Chloroplast two-component systems: evolution of the link between photosynthesis and gene expression. *Proc. Biol. Sci., 276*, 2133–2145.

Puthiyaveetil, S., Kavanagh, T. A., Cain, P., Sullivan, J. A., Newell, C. A., Gray, J. C., Robinson, C., van der Giezen, M., Rogers, M. B., & Allen, J. F. (2008). The ancestral symbiont sensor kinase CSK links photosynthesis with gene expression in chloroplasts. *Proc. Natl. Acad. Sci. USA, 105*, 10061–10066.

Raven, P., Johnson, G., Mason, K., Losos, J., & Singer, S. (2010). *Biology* (9th ed.). New York: McGraw-Hill.

Reece, R. J., & Maxwell, A. (1991). DNA gyrase: structure and function. *Crit. Rev. Biochem. Mol. Biol., 26*, 335–375.

Rich, P. R. (2003). The molecular machinery of Keilin's respiratory chain. *Biochem. Soc. Trans., 31*, 1095–1105.

Schneider, R., Travers, A., & Muskhelishvili, G. (2000). The expression of the *Escherichia coli* fis gene is strongly dependent on the superhelical density of DNA. *Mol. Microbiol., 38*, 167–175.

Stoimenova, M., Igamberdiev, A. U., Gupta, K. J., & Hill, R. D. (2007). Nitrite-driven anaerobic ATP synthesis in barley and rice root mitochondria. *Planta, 226*, 465–474.

Taylor, S. W., Fahy, E., Zhang, B., Glenn, G. M., Warnock, D. E., Wiley, S., Murphy, A. N., Gaucher, S. P., Capaldi, R. A., Gibson, B. W., & Ghosh, S. S. (2003). Characterization of the human heart mitochondrial proteome. *Nat. Biotechnol., 21*, 281–286.

Tseh-Dinh, Y. C., & Beran, R. K. (1988). Multiple promoters for transcription of the *Escherichia coli* DNA topoisomerase I gene and their regulation by DNA supercoiling. *J. Mol. Biol., 202*, 735–742.

Voet, D., Voet, J. G., & Pratt, C. W. (2006). *Fundamentals of biochemistry* (2nd ed.). New York: John Wiley and Sons.

Westerhoff, H. V., O'Dea, M. H., Maxwell, A., & Gellert, M. (1988). DNA supercoiling by DNA gyrase. A static head analysis. *Cell Biophys., 12*, 157–181.

Whitehall, S., Austin, S., & Dixon, R. (1993). The function of the upstream region of the sigma 54-dependent *Klebsiella pneumoniae* nifL promoter is sensitive to DNA supercoiling. *Mol. Microbiol., 9*, 1107–1117.

Zhang, J., Li, X., Mueller, M., Wang, Y., Zong, C., Deng, N., Vondriska, T. M., Liem, D. A., Yang, J., Korge, P., Honda, H., Weiss, J. N., Apweiler, R., & Ping, P. (2008a). Systematic characterization of the murine mitochondrial proteome using functionally validated cardiac mitochondria. *Proteomics, 8*, 1564–1575.

Zhang, J., Liem, D. A., Mueller, M., Wang, Y., Zong, C., Deng, N., Vondriska, T. M., Yang, J., Korge, P., Drews, O., Maclellan, W. R., Honda, H., Weiss, J. N., Apweiler, R., & Ping, P. (2008b). Altered proteome biology of cardiac mitochondria under stress conditions. *J. Proteome Res., 7*, 2204–2214.

Zhu, Y. S., & Hearst, J. E. (1988). Transcription of oxygen-regulated photosynthetic genes requires DNA gyrase in *Rhodobacter capsulatus*. *Proc. Natl. Acad. Sci. USA, 85*, 4209–4213.

CHAPTER

Towards Direct Electrochemical Control of Gene Expression Level

7

CHAPTER OUTLINE

7.0 Direct electrochemical control of gene expression level: where to start from?

In the previous chapter we have considered a number of cases, involving different bacteria and subcellular organelles in eukaryotes, of redox-dependent modulation of gene expression pathways. Albeit the general picture of the modulation of gene expression is very complex and mostly not fully elucidated yet, the case of photosynthetic bacteria appears sufficiently clear to select it as a possible candidate for technological exploitation. We are, indeed, interested in developing approaches to the modulation of gene expression by an external stimulus, in order to gain high-level control of the phenomenon. The ultimate goal of this activity deals, of course, with the possibility of making a biosystem amenable to our needs, following a track resembling the one that has led us to exploit bacteria, and cells in general, as protein factories for biotechnological purposes. Approaches that tend toward a similar goal are, for example, those developed by *optogenetics*, which is a neuromodulation technique employed in neuroscience; it uses a combination of techniques from optics and genetics to control the activity of individual neurons in living tissue – even within freely-moving animals – and to precisely measure the effects of those manipulations in real time (Deisseroth et al., 2006).

The interest in redox-regulated gene expression stems from the notions and the expertise we have gained in all the previous chapters, where we

Biomolecular Electronics. http://dx.doi.org/10.1016/B978-1-4557-3142-8.00007-8

have learnt how to control a working electrode potential, to drive the redox state of immobilized and freely diffusing molecules, etc.

Now we are in a position to try to design a possible electrical approach to the control of gene expression, but, to achieve such an ambitious goal, we first have to select a suitable biological system. A plausible choice appears to be to focus on photosynthetic bacteria of the genus *Rhodobacter.* The reasons for this choice are connected with a number of interesting and suitable features that one can find in these bacteria, apart from the prerequisite of featuring redox-dependent gene expression molecular machinery. Indeed, as we have seen in section 6.2, variation in oxygen tension is responsible for modulating the oxidative state of the inner bacterial compartment that activates the change of metabolism in the microorganism and makes it a facultative photosynthetic bacterium.

The ability that these bacteria have to switch from photosynthesis to respiration is at the core of a straightforward way to monitor the products of their genes. Indeed, the photosynthetic state (low oxygen tension) tends to enhance expression of those genes that encode for proteins involved in light harvesting and in light-induced charge separation (i.e., LHI, LHII, reaction centers), along with the connected pigments, whereas high oxygen tension tends to depress their expression. Therefore, both *R. sphaeroides* and *R. capsulatus* (see section 6.2) are convenient objects to monitor in their metabolic variation, simply by following the variation of the optical absorption bands related to their pigments (bacteriochlorophyll monomers and dimers, bacteriopheophytins) brought about by the proteins of photosynthesis.

Other positive aspects of the genus at issue are its robustness and relative ease of growth, which make them good candidates for repeated experimentation in non-conventional environments.

Affecting the inner redox state of a bacterium by direct electrochemistry implies a “hot-wiring” of the bacterial cytoplasm with an external voltage source, e.g., a potentiostat. A direct physical connection is, however, difficult to implement and doable only one cell at a time; as such it appears of limited technological interest.

We seek, indeed, an approach suitable to affect, contemporarily, the inner redox state of a large bacterial population. A plausible possibility to achieve this goal is to use redox molecular relays or mediators whose oxidation state can be conditioned by the interaction with an electrode in the culture medium, whose potential is driven by a potentiostat. These mediators have to be able to freely diffuse across the bacterial membrane, penetrating the cytoplasm and affecting its redox state. By doing so, they play the role of a metal wire piercing the bacterial membrane and directly affecting the redox state inside the cell.

The role of these mediators has to be that of mimicking the effect of oxygen tension variations in modulating gene expression. Therefore, as we have seen in section 6.2, they have to be able to affect the redox state of thioredoxins that in turn differentially bind gyrases, resulting in a different supercoiling of the bacterial chromosome, hence in a modulation of the transcription of the photosynthesis genes.

The overall project appears quite ambitious and its various steps are best faced if we try to dissect the process into its basic constituents. For each step or phase, a specific approach will be sought and solutions will be suggested often using a crucial interplay between top-down and bottom-up approaches. In the next sections we will analyze in detail the principal main phases of a process whose completion can lead to implementing the sought electrical control.

7.1 The choice of redox mediators

Which are the features that a redox mediator should have in order to accomplish the task of affecting the redox state of thioredoxins in the desired way? There are, indeed, several aspects to take into account in order to come up with a satisfying choice.

Low-molecular-weight redox species can assist the shuttling of electrons between the intracellular bacterial space and an electrode. However, there are many important requirements that such a mediator should satisfy in order to provide efficient electron transport from/to the bacterial thioredoxins to an external electrode: (a) The mediator should easily penetrate through the bacterial membrane to reach the thioredoxins inside the bacteria. (b) The redox-potential of the mediator should fit the potential of the thioredoxins (e.g., the mediator potential should be positive enough to provide fast electron transfer from the enzymes, but it should not be too positive as to prevent significant loss of potential). (c) Neither oxidation state of the mediator should interfere with other metabolic processes (should not inhibit them or be decomposed by them). (d) The mediator should easily escape from the cell through the bacterial membrane. (e) The oxidation states of the mediator should be chemically stable in the electrolyte solution, they should be well soluble, and they should not adsorb onto the bacterial cells or electrode surface. (f) The electrochemical kinetics of the redox process of the mediator at the electrode should be fast (electrochemically reversible) (Wilkinson, 2000).

Luckily enough, one can take advantage of the extended screening activity on these kinds of redox mediators, conducted in studies dedicated to design and implementation of bacterial fuel cells (Katz et al., 2003).

Many different organic and organometallic compounds have been tested in combination with bacteria to evaluate the efficiency of mediated electron transport from the internal bacterial metabolites to the anode of a biofuel cell. Thionine has been used extensively as a mediator of the electron transport from *Proteus vulgaris* (Bennetto et al., 1985; Thurston et al., 1985; Delaney et al., 1984; Kim et al., 2000) and from *E. coli* (Roller et al., 1984; Bennetto et al., 1983). Other organic dyes that have been used include benzylviologen, 2,6-dichlorophenolindophenol, 2-hydroxy-1,4-naphthoquinone, phenazines (phenazine ethosulfate, safranine), phenothiazines (alizarine brilliant blue, *N,N*-dimethyl disulfonated thionine, methylene blue, phenothiazine, toluidine blue) and phenoxazines (brilliant cresyl blue, gallocyanine, resorufin) (Delaney et al., 1984; Roller et al., 1984; Bennetto et al., 1983; Park & Zeikus, 2000; Ardeleanu et al., 1983; Davis & Yarbrough, 1962; Patchett et al., 1988; Kreysa et al., 1990). These organic dyes were tested with *Alcaligenes eutrophus*, *Anacystis nidulans*, *Azotobacter chroococcum*, *Bacillus subtilis*, *Clostridium butyricum*, *E. coli*, *Proteus vulgaris*, *Pseudomonas aeruginosa*, *Pseudomonas putida* and *Staphylococcus aureus* bacteria, usually using glucose and succinate as substrates. Among the evaluated dyes, phenoxazine, phenothiazine, phenazine, indophenol, bipyridilium derivatives, thionine and 2-hydroxy-1,4-naphthoquinone were found to be very efficient in maintaining relatively high cell voltage output when current was drawn from the biofuel cell (Delaney et al. 1984). Some other dyes do not function as effective mediators because they are not rapidly reduced by the microorganisms, or they lack sufficient negative potential. Ferric chelate complexes (e.g., Fe(III)EDTA) were successfully used with *Lactobacillus plantarum*, *Streptococcus lactis* and *Erwinia dissolvens*, oxidizing glucose (Vega & Fernández, 1987).

Since thionine has frequently been used as a mediator in microbial fuel cells, mono- and disulfonated derivatives of thionine have been applied to determine the effect of hydrophilic substituents on mediation of electron transfer from *E. coli* to an anode (Lithgow et al., 1986). Changing from thionine to 2-sulfonated thionine and 2,6-disulfonated thionine results in an increased efficiency of mediated electron transport. The low efficiencies of the biofuel cells operating with thionine and 2-sulfonated thionine were attributed to interference with electron transfer by adsorption of the mediator on the microbial membrane. It should be noted that the overall efficiency of the electron-transfer mediators depends also on many other parameters, and in particular on the electrochemical rate constant of mediator re-oxidation, which depend, in turn, on the electrode material.

Since an electron-transfer mediator needs to meet many requirements, some of which are mutually exclusive, it is not possible to reach perfect conditions for electron transport from a bacterial cell to an electrode. A mixture of two mediators can be useful in optimizing efficiency. A solution containing thionine and Fe(III)EDTA was applied to mediate electron transport from *E. coli*, oxidizing glucose as a primary substrate to an anode (Tanaka et al., 1983). Although both mediators can be reduced by *E. coli*, thionine is reduced over 100 times faster than Fe(III)EDTA. The electrochemical oxidation of the reduced thionine is much slower than oxidation of Fe(II)EDTA, however. Therefore, electrons obtained from the oxidation of glucose in the presence of *E. coli* are transferred mainly to thionine under the operational conditions of the cell. The reduced thionine is rapidly re-oxidized by Fe(III)EDTA, the rate of which has been shown to be very fast, $k_{et} = 4.8\times10^4\ M^{-1}s^{-1}$. Finally, the reduced chelate complex, Fe(II)EDTA, transfers electrons to the anode by the electrode reaction of a Fe(III)EDTA/Fe(II)EDTA couple with a sufficiently large rate constant. One more example of enhanced electron transport in the presence of a mixture of mediators was shown for *Bacillus* strains oxidizing glucose as a primary substrate. The biofuel cell was operated in the presence of methylviologen (MV^{2+}) and 2-hydroxy-1,4-naphthoquinone or Fe(III) EDTA (Akiba et al., 1985). Methylviologen can efficiently accept electrons from the bacterial cells, but its reduced state ($MV^{\bullet+}$) is highly toxic for the bacteria and immediately inhibits the fermentation process. In the presence of a secondary mediator that has a more positive potential, $MV^{\bullet+}$ is efficiently re-oxidized to MV^{2+}. The reduced secondary mediator (quinone or Fe(II) EDTA) then transports the electrons to the anode.

In Figure 7.1 the relative typical redox potentials of thioredoxin (−0.47 V vs SCE) (Watson et al., 2003) along with those of safranin (−0.53 mV) and

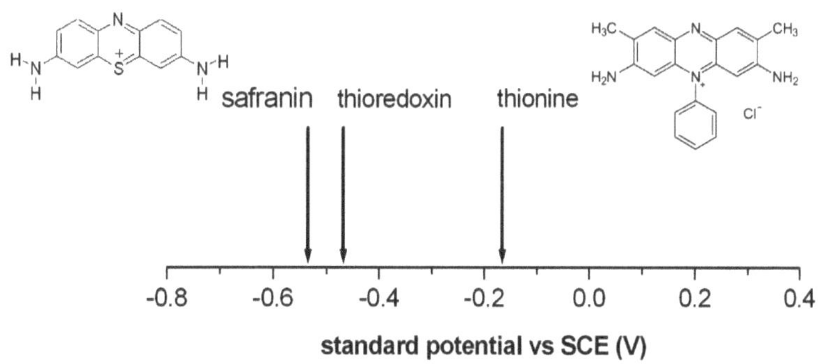

FIGURE 7.1

The relative positions of the standard potential of thioredoxins, safranin, and thionine are indicated by the arrows. The structures of the two dyes are given beside the corresponding arrows.

thionine (0.18 mV) are reported with the structure of both redox mediators. These two dyes appear to be suitable for affecting the redox state of thioredoxin. Particularly, whereas safranin can reduce the enzyme, thionine can oxidize it. Therefore, these two molecules seem to be a suitable choice for controlling the redox state of the disulfides of thioredoxins. We know, from section 6.2, that the redox state of thioredoxin's S-S bridges determines whether the molecule is able to bind gyrase, determining eventually the fate of the gene expression in *Rhodobacter.*

7.2 How to go further?

After having selected a suitable pair of redox mediators that could in principle accomplish the task of affecting the redox state of thioredoxin's disulfides in the desired way, one has to face the question of how to proceed in tackling the overall problem. Indeed, the variables to be controlled are so numerous that it is difficult to proceed immediately towards a real biological system if one aims at understanding what is going on. It is of course possible to try to evaluate whether a serendipitous approach can provide any positive results, but a detailed understanding and control of the various steps that can lead to gene expression control definitely require the implementation of a systematic study.

Such an all-or-none attempt has indeed been done using *Rhodobacter sphaeroides*. The experiment compared the absorption spectra of two different bacterial suspensions in the bacteriochlorophyll spectral range. One of these was left to grow for 24 h in normal aerobic conditions, whereas the second was incubated with 1 mM thionine. The flask also contained three electrodes, a Au wire as WE, a reference (SCE) and a counter (Pt). The working electrode was kept at a potential of 0 V for 24 h. Such a potential helped maintain thionine in an oxidized state. Oxidized thionine, once diffusing inside bacteria, could reduce the disulfides of thioredoxins, providing an increase in the supercoiling activity of gyrases on the bacterial chromosome that should cause a decrease in *puc* and *puf* operon transcription (see section 6.2), hence a reduced amount of bacteriochlorophyll present in the bacteria. After 24 h, the amount of bacteria in the two flasks was evaluated by optical absorption spectroscopy; subsequently, bacterial concentration was adjusted to a value suitable for absorption spectroscopy in the bacteriochlorophyll spectral range (700–900 nm) and the corresponding spectra were acquired in a comparative fashion. Figure 7.2 shows the spectra obtained, which show a decrease in the normalized bacteriochlorophyll absorption of the sample under potentiostatic control, whereas the control, without the redox

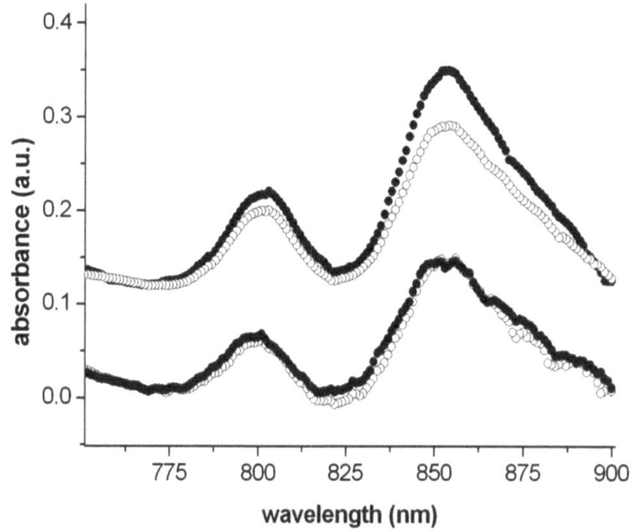

FIGURE 7.2

Absorption spectra of the two different bacterial suspensions exposed (black dots) and not exposed (white circles) to direct electrochemical control of the cytoplasmic oxidation state. The lower spectra are acquired on samples without the presence of thionine but applying a potential to the flask (black dots) and in the absence of both thionine and electrochemical control.

mediator, does not show any appreciable difference due to the application of a potential.

These data are just preliminary and, as such, they do not allow one to draw any robust conclusions on the validity of the approach used. Of course, the questions still to be answered are numerous and even the positive identification of the mechanisms at play can be questioned (e.g., whether the redox mediator operates according to the described mechanism, see section 6.2, or whether it directly affects the bacteriochlorophyll oxidation state).

We believe that to go any further in the mechanistic understanding of the phenomena taking place during redox control of the cytoplasmic space by direct electrochemistry, one needs a more systematic approach probably involving the dissection of the problem into its main constituent parts. At any rate, these preliminary results appear to be promising and deserve further attention.

It is now useful to think of possible ways to analyze the process by subdividing it into a number of steps that allow one to achieve full control of the phenomenon. One can think that the first step to overcome is to demonstrate that the proposed approach is really able to control the oxidation state of the inner space of a model system mimicking a cell envelope. It is conceivable to have an artificial lipid envelope filled with a suitable redox molecule whose oxidation state could be affected by the use of the selected redox mediator. A possibility could be to fill the envelope with potassium ferricyanide, which possesses a redox potential ≈ 436 mV at pH 7, and which could be easily reduced to ferrocyanide by thionine or safranin according to: $[Fe(CN)_6]^{4-} \rightleftharpoons [Fe(CN)_6]^{3-} + e^-$. This reaction can be followed spectroscopically (at 420 nm, molar extinction coefficient = 1040 $M^{-1}cm^{-1}$) or optically (under a microscope and with suitable concentrations, the reduction reaction makes the dye change its color from brown to pale yellow). Furthermore both ferro- and ferricyanide are known to be impermeable through the plasma membrane, thus ensuring their confinement inside the envelope.

Quite natural candidates for this kind of test appear to be liposomes. They are one of the several possible self-assembling structures that amphiphilic molecules such as phospholipids can form when dispersed in an aqueous solution. For a comprehensive exposition of these and other self-assembling structures the reader can refer to any of the numerous reviews and books present in the literature (for a quantitative description see Israelachvili, 2005). These aggregates can be formed in various sizes in the range from a few nm to several tens of micrometers. According to their typical size, they are called SUVs (for Small Unilamellar Vesicles, from a few to tens of nanometers), LUVs (for Large Unilamellar Vesicles, from tens to hundreds of nanometers), or GUVs (for Giant Unilamellar Vesicles, in the micron to tens of microns range). They consist of a single phospholipid bilayer that, for thermodynamic reasons connected to the minimization of the free energy associated with the unsaturated boundaries, bends and forms a closed envelope that mimics a biological membrane. Considering that it is possible to assemble liposomes from lipid mixtures that are present in bacterial membranes (e.g., PE:PG in 3:1 molar ratio is found in the inner *E. coli* membrane), these structures appear to be suitable candidates for mimicking their behavior with regard to the main aspect of our interest, which is the possibility of affecting their inner redox state by redox mediators.

7.2.1 LUVs and GUVs

The relative ease of preparation of self-assembling phospholipid structures along with the existence of suitable experimental methods to probe phenomena taking place in their interior make LUVs and GUVs systems of choice for implementing the early steps towards a mechanistic understanding of the phenomena behind the direct electrochemical control of the redox environment inside cells.

LUVs and GUVs have a fundamental difference in that the former are amenable to optical spectroscopic, bulk investigation (e.g., UV-Vis optical absorption spectroscopy), whereas the latter, given their quite large size, can be studied and observed under an optical microscope (e.g., fluorescence microscopy). Therefore we propose both of these systems as suitable bench tests for tracking changes in the oxidation state of their interior.

Preparations of liposomes of different sizes share a number of common steps. When dealing with liposomes of mixed lipid composition, the lipids must first be dissolved and mixed in an organic solvent to ensure a homogeneous mixture is achieved. Usually this process is carried out using chloroform or $CHCl_3$/MetOH mixtures (according to lipid solubility). One has to obtain a clear lipid solution for complete mixing of the lipids. Typically, lipid solutions are prepared at 10–20 mg lipid/ml organic solvent. Once the lipids are thoroughly mixed in the organic solvent, the solvent is removed to yield a lipid film. In the case of small solvent volumes (< 1 ml), the solvent can be evaporated using a dry nitrogen or argon stream in a fume hood. For larger volumes, the organic solvent should be removed by rotary evaporation, yielding a thin lipid film on the sides of a round-bottom flask. The lipid film is thoroughly dried to remove residual organic solvent by placing the vial under vacuum for a variable time (usually several hours). Dry lipid films can then be stored frozen until ready to hydrate.

Hydration of dry lipid films can be achieved by simply adding an aqueous medium to the container of dry lipids and agitating. The temperature of the hydrating medium should be above the gel–liquid crystal transition temperature (T_c or T_m) of the lipid with the highest T_c before adding to the dry lipid. After adding the hydrating medium, the lipid suspension should be kept above the T_c during the whole period. The hydration medium is generally determined by the application of the lipid vesicles. Suitable hydration media typically include distilled water, buffer solutions, saline, and non-electrolytes such as sugar solutions.

Typical hydrating solution compositions that meet these conditions are 0.9% saline, 5% dextrose and 10% sucrose. During hydration some lipids form complexes unique to their structure. The results of the described hydration procedure are large multilamellar vesicles (LMVs) whose structure resembles an onion, with each lipid bilayer separated by a water layer. Once a stable, hydrated LMV suspension has been achieved, the particles can be downsized by a variety of techniques, including sonication or extrusion.

Disruption of LMV suspensions using sonication typically gives rise to small unilamellar vesicles (SUV) with diameters in the range of 15–50 nm that can also be useful in some cases for our aims. The most common techniques for preparation of such particles are bath and probe tip sonicators. Bath sonicators, due to their limited power and higher level of cleanness, are the most widely used instrumentation for preparation of SUV. Sonication of an LMV dispersion is achieved by placing a test tube containing the suspension in a bath for 5–10 minutes above the T_c of the lipid. As a result, the lipid suspension begins to clarify to yield a slightly hazy transparent solution. The achievable mean size and distribution of the SUVs are influenced by composition and concentration, temperature, sonication time and power, and volume. Furthermore, due to the high degree of curvature of these membranes, SUVs are inherently unstable and spontaneously fuse to form larger vesicles when stored below their phase-transition temperature.

Lipid extrusion is the approach of choice to yield LUVs. It consists in forcing a lipid suspension through a polycarbonate filter with a defined pore size to yield particles having a diameter dictated by the pore size of the filter used. This procedure has to be repeated typically around 10 times, to achieve repeatable final results. Prior to extrusion through the final pore size, LMV suspensions are disrupted either by several freeze–thaw cycles or by pre-filtering the suspension through a larger pore size (typically 0.2–1.0 μm). This method helps prevent the membranes from fouling and improves the homogeneity of the size distribution of the final suspension. As with all procedures for downsizing LMV dispersions, the extrusion should be performed at a temperature above the T_c of the lipid. Extrusion through filters with 100 nm pores typically yields large unilamellar vesicles (LUV) with a mean diameter of 120–140 nm. Mean particle size also depends on lipid composition and is quite reproducible from batch to batch. In Figure 7.3 a typical extruding set-up is shown.

In the specific case at issue, LUVs have to be extruded in a solution also containing the target molecule that is going to be used as an indicator of the redox state inside the liposome. In the present case 1 mg/ml potassium ferricyanide could be enough. LUVs eventually have to be dialyzed against a suitable (physiological) buffer in order to get rid of external ferricyanide from the solution.

Preparation of GUVs can be performed following different approaches. One of the most reliable is to use a dedicated growth chamber where two Pt wires (typically 0.5 mm in diameter) are electrically biased with a slowly varying sinusoidal potential (a few volts p-p and a frequency in the range 2–10 Hz) after having been painted with phospholipid solution. According to the different types of phospholipids used, solvent composition can also

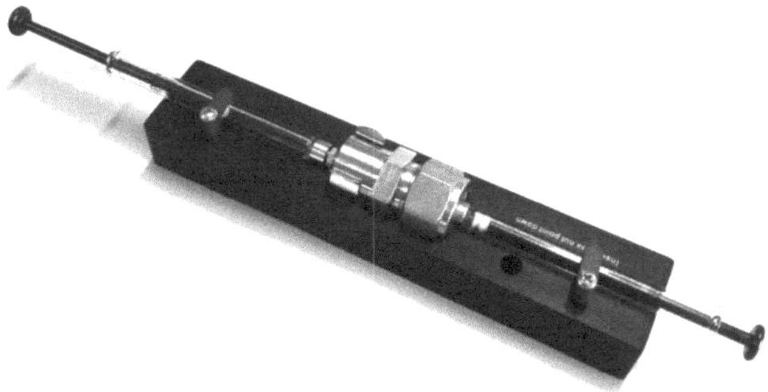

FIGURE 7.3

A typical extruding device for LUV production.

vary. In the case of a PC headgroup, as reported in the following example, 1 mg/ml DOPC in $CHCl_3$/MetOH (50%/50% vol.) has been used. After proper drying, the growth process is to be carried out in a water-based solution with a composition that depends upon the particular lipids used and their foreseen application but that usually contains sucrose (200 mM) and a salt at quite low ionic strength ($<$ 10 mM).

Figure 7.4 shows a typical image of a GUV attached to a suction pipette, while the inset shows a GUV filled with methylene blue.

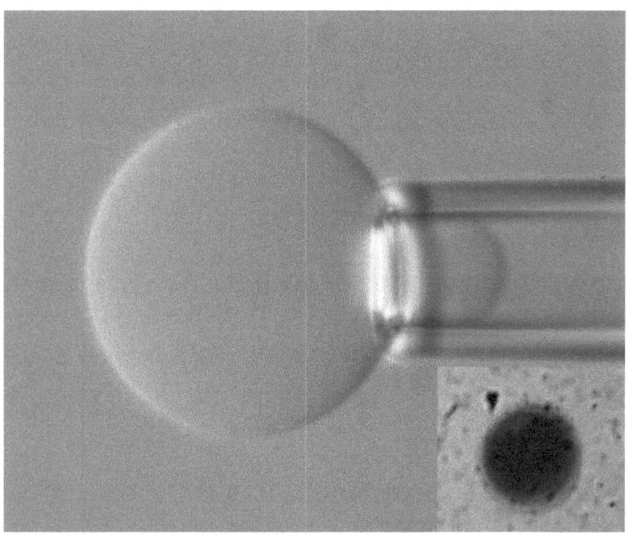

FIGURE 7.4

A 60 nm large GUV made of DOPC sucked by a micropipette. Inset: a similar GUV prepared in 10^{-4} M methylene blue.

It is therefore suggested that the proposed systems can be valuable to test the effect of redox mediators in influencing the oxidation state of the inner compartment of lipid vesicles. Measurements on LUVs and GUVs will enable one to quantify the kinetics of the phenomenon, as well as the critical concentration and potential values to be used in order to achieve the desired result.

7.2.2 Complicating the system

If one wishes to go further along the steps needed to attain complete control over the electrochemical modulation of the gene expression level in photosynthetic bacteria, one has also to implement a strategy suitable to check whether the electrochemical conditioning of the inner vesicle space can really cause the foreseen cascade of events (see section 6.2).

Therefore, one can continue with the vesicle approach, complicating the system. A possible way of doing that is to make use of recombinant enzymes that can more closely resemble the real, biological system that remains the final goal of the experimentation. The availability of recombinant thioredoxins and DNA gyrases allows one to think of an artificial system where there are present all the main constituents of the foreseen cascade of molecular events that should lead to the modulation of the supercoiling of the bacterial chromosome, hence driving the access of DNA polymerases to different portions of the genome.

The proposal foresees therefore the formation of liposomes in the presence of thioredoxin, gyrase, ATP and circular DNA (e.g., circular λ-DNA or plasmids). Circular DNA supercoiling can be checked by gel-electrophoresis of the pristine DNA. Once all the molecular components are ready, it will be necessary to form liposomes containing all of them in the correct relative concentrations. Submitting the loaded liposomes to redox state control with either safranin or thionine, the modulation of the oxidation state of the S-S bridges of thioredoxin will be achieved, providing as a consequence the cascade of events outlined in section 6.2 and schematically summarized in Figure 6.1.

After a suitable conditioning time, the recovery of DNA and the analysis of its supercoiling (by electrophoresis or even by atomic force microscopy) will provide a quite direct answer to the question about the efficacy of the proposed approach.

In particular, decreased supercoiling of the recovered DNA (Nelson & Cox 2008) with respect to that inserted in the liposomes will be direct evidence of the fact that the control of the SS vs the SH state in thioredoxin has been effectively achieved by direct electrochemistry. The proposed synthetic approach has the obvious advantage of allowing for a detailed screening of the various phases constituting the logical development of the process of DNA-supercoiling modulation. Once achieved, the described result will enable a much safer belief in the results that have been preliminarily shown in Figure 7.2 and that of course will be amenable to optimization in light of the outcome of the "synthetic" experimental approach hereby proposed.

7.3 References

Akiba, T., Bennetto, H. P., Stirling, J. L., & Tanaka, K. (1985). Electricity generation from alkalophilic organisms. *Biotechnol. Lett., 9*, 611–616.

Ardeleanu, I., Margineanu, D. G., & Vais, H. (1983). Electrochemical conversion in biofuel cells using *Clostridium butyricum* or *Staphylococcus aureus. Bioelectrochem. Bioenerg., 11*, 273–277.

Bennetto, H. P., Stirling, J. L., Tanaka, K., & Vega, C. A. (1983). Anodic reactions in microbial fuel cells. *Biotechnol. Bioeng., 25*, 559–568.

Bennetto, H. P., Delaney, G. M., Mason, J. R., Roller, S. D., Stirling, J. L., & Thurston, C. F. (1985). The sucrose fuel cell: efficient biomass conversion using a microbial catalyst. *Biotechnol. Lett., 7*, 699–704.

Davis, J. B., & Yarbrough, H. F., Jr. (1962). Preliminary experiments on a microbial fuel cell. *Science, 137*, 615–616.

Deisseroth, K., Feng, G., Majewska, A. K., Miesenbock, G., Ting, A., & Schnitzer, M. J. (2006). Next-generation optical technologies for illuminating genetically targeted brain circuits. *J. Neurosci., 26*, 10380–10386.

Delaney, G. M., Bennetto, H. P., Mason, J. R., Roller, S. D., Stirling, J. L., & Thurston, C. F. (1984). Electron-transfer coupling in microbial fuel cells; 2. Performance of fuel cells containing selected microorganism-mediator-substrate combination. *J. Chem. Technol. Biotechnol., 34B*, 13–27.

Israelachvili, J. N. (2005). *Intermolecular and surface forces* (3rd ed.). New York: Academic Press.

Katz, E., Shipway, A. N., & Willner, I. (2003). Biochemical fuel cells. In W. Vielstich, H. A. Gasteiger, & A. Lamm (Eds.), *Handbook of fuel cells fundamentals, technology and applications* (vol. 1); (pp. 355–382). London: John Wiley and Sons.

Kim, N., Choi, Y., Jung, S., & Kim, S. (2000). Effect of initial carbon sources on the performance of microbial fuel cells containing Proteus vulgaris. *Biotechnol. Bioeng., 70*, 109–114.

Kreysa, G., Sell, D., & Krämer, P. (1990). Bioelectrochemical fuel cells. *Ber. Bunsenges. Phys. Chem., 94*, 1042–1045.

Lithgow, A. M., Romero, L., Sanchez, I. C., Souto, F. A., & Vega, C. A. (1986). Interception of electron-transport chain in bacteria with hydrophilic redox mediators. *J. Chem. Res. Synop., 5*, 178–179.

Nelson, D. L., & Cox, M. M. (2008). *Lehninger principles of biochemistry* (5th ed.). New York: W. H. Freeman.

Park, D. H., & Zeikus, J. G. (2000). Electricity generation in microbial fuel cells using neutral red as an electronophore. *Appl. Environ. Microbiol., 66*, 1292–1297.

Patchett, R. A., Kelly, A. F., & Kroll, R. G. (1988). Use of a microbial fuel cell for the rapid enumeration of bacteria. *Appl. Microbiol. Biotechnol., 28*, 26–31.

Roller, S. D., Bennetto, H. P., Delaney, G. M., Mason, J. R., Stirling, S. L., & Thurston, C. F. (1984). Electron-transfer coupling in microbial fuel cells: 1. Comparison of redox mediator reduction rates and respiration rates of bacteria. *J. Chem. Technol. Biotechnol., 34B*, 3–12.

Tanaka, K., Vega, C. A., & Tamamushi, R. (1983). Thionine and ferric chelate compounds as coupled mediators in microbial fuel cells. *Bioelectrochem. Bioenerg., 11*, 289–297.

Thurston, C. F., Bennetto, H. P., Delaney, G. M., Mason, J. R., Roller, S. D., & Stirling, J. L. (1985). Glucose metabolism in a microbial fuel cell. Stoichiometry of product formation in a thionine-mediated *Proteus vulgaris* fuel cell and its relation to coulombic yields. *J. Gen. Microbiol., 131*, 1393–1401.

Vega, C. A., & Fernández, I. (1987). Mediating effect of ferric chelate compounds in microbial fuel cells with *Lactobacillus planctarum, Streptococcus lactis*, and Erwinia dissolvens. *Bioelectrochem. Bioenerg., 17*, 217–222.

Watson, W. H., Pohl, J., Montfort, W. R., Stuchlik, O., Reed, M. S., Powis, G., & Jones, D. P. (2003). Redox potential of human thioredoxin 1 and identification of a second dithiol/disulfide motif. *J. Biol. Chem., 278*, 33408–33415.

Wilkinson, S. (2000). Gastrobots – benefits & challenges of microbial fuel cells in food powered robot applications. *J. Autonomous Robots, 9*, 99–111.

CHAPTER

What will be Next?

8

CHAPTER OUTLINE

8.0 A pervasive presence of redox-controlled biosystems

The largest majority of relevant biological reactions and systems that are prone to be directly or indirectly affected by the action of electric fields are, as we have seen, quite complex and often still far from being fully unraveled both in their mechanistic aspects and in the number of consequences and implications that each of them brings about.

Redox biology, that branch of biology that focuses on understanding phenomena involving changes in the oxidation state of molecular partners, is constantly expanding the number of relevant cases that appear to be redox-regulated; moreover, these cases belong to an increasing variety of biological situations and contexts, ranging from cellular energetics, to the control of gene expression level, plant growth and development, stem cell differentiation, blood pressure, activity of p53, etc., just to quote some of the most relevant cases. As the understanding of these numerous and diverse phenomena deepens, it is very likely that the possibility of implementing technological control over them will also develop in turn.

In the next sections we will present some of these cases that appear particularly promising to us. It is, however, our firm conviction that future

Biomolecular Electronics. http://dx.doi.org/10.1016/B978-1-4557-3142-8.00008-X

developments in science and technology will provide many more opportunities and exploitable cases on this topic than we can now even imagine.

8.1 Redox-dependent control of blood pressure

Blood pressure is regulated by many signals arising from the kidneys, blood vessels, heart and brain and any imbalance among them can give rise to altered blood-pressure values. One of these signaling pathways is that realized by the renin-angiotensin system. The enzyme renin is a protease that catalyzes the cleavage of the 10-amino-acid terminus at the N-terminus of its substrate, angiotensinogen. The result of the cleavage step is further processed to give rise to angiotensin peptide hormones, among which the best characterized is angiotensin II, whose effect is to increase blood pressure in various ways. Even if this hormone can activate a variety of cellular processes while binding to its cellular receptors, a shared aspect is the formation of reactive oxygen species (ROS) (Rajagopalan et al., 1996). As an example, the activation of angiotensin receptors can induce the production of superoxide and cause oxidative stress in the main cardiovascular target organs such as the kidneys, blood vessels and brain. Superoxide represents indeed a highly reactive free radical that can be converted to lipid-soluble products such as hydrogen peroxide and peroxynitrite. It has been shown in rat and mouse models of angiotensin-induced hypertension that treatment with antioxidant can decrease blood pressure, thus providing further support for the hypertensive effects of ROS as generated by the renin-angiotensin system.

One of the effects of oxidative stress is its action on cysteine amino acid residues in proteins. For example, interaction between H_2O_2 and the thiol group of cysteines yields the formation of a sulfenic acid intermediate that can react with cysteines, giving rise to disulfide bonds that can alter protein structure and function. This kind of oxidative protein alteration appears to play a role also in the control of hypertension. Indeed, angiotensinogen's amino terminal tail, along with the angiotensin cleavage site, also undergoes a conformational rearrangement that makes it accessible to renin. The essential aspect of this conformational change seems to be the formation of a disulfide bridge between a pair of cysteines (C18 and C138) of angiotensinogen. The oxidized state of this protein correlates with its increased affinity for renin, hence, with an enhanced production of angiotensin that results in increased hypertension (Zhou et al., 2010). Interestingly,the C18-C138 disulfide appears to be redox sensitive, being reducible by the action of glutathione, which regenerates two unbound cysteine residues. This fact can have important physiological implications since endogenous nitric oxide, which is a potent dilator of blood vessels, hence reducing blood pressure, can react with cysteine thiols by *S*-nitosylation (addition of NO group), altering their chemical nature. Indeed, it has been found that nitric oxide can react with and block the thiols of C18 and C138, possibly preventing further oxidation of the two cysteines.

These findings point to a tight regulation of the redox state of angiotensinogen in tissues by intracellular chemicals such as ROS and nitric oxide. As a conclusion, the rate of angiotensin formation can be determined by the redox state of tissues and blood and by the absolute amount of angiotensinogen and renin present.

The described results suggest this redox biosystem as a potential target for direct electrochemical control and open up the way for considering this kind of approach also for application as a preventive/therapeutic means in humans.

8.2 Redox regulation of embryonic stem cell transcription factors by thioredoxin

We have already met the key role of bacterial thioredoxins in ruling the redox-dependent mechanism of gene expression regulation in photosynthetic bacteria (see section 6.2). Thioredoxins play a prominent role also in the control of gene expression in eukaryotic organisms and hereby an example is reported of their key role in regulating embryonic stem cell transcription factors.

Also in eukaryotic organisms some transcription factors are regulated by variations in environmental redox state. This is the case, for instance, for PEBP2, AP-1, p53, NF-κB. Their redox regulation occurs through conserved cysteine residues in the DNA-binding domains of these proteins (Hirota et al., 1997, 1999; Ueno et al., 1999; Akamatsu et al. 1997). Two intracellular enzymes exist that can restore the DNA-binding ability of oxidized transcriptional regulators by reducing critical cysteines: thioredoxin and APE-1/Ref-1.

Also lineage commitment and stage progression during embryonic stem cell development is controlled by transcription factors. Particularly, Oct-4 is one of the regulators, required to maintain the totipotentiality of embryonic stem cells. Indeed, down-regulation of its activity is required for proper differentiation of the blastocyst during uterine implantation. Uterine implantation and subsequent vascularization increase the developing embryo's exposure to oxygen, altering, as a consequence, the intracellular redox state.

Oct-4 activity can be regulated by these changes in redox state. Particularly, oxidation abolishes Oct-4 binding to DNA, at variance with what happens in the case of another embryonic cell transcription factor – FoxD3. The role of thioredoxin appears ideally suited to restore the DNA-binding activity of Oct-4, whereas it is much less effective in restoring that of FoxD3. A mechanistic understanding of this fact traces the reasons for this difference back to the ability of thioredoxin (but not Ape-1) to associate with cysteines in the POU domain of Oct-4 (Guo et al., 2004). This fact indicates that there may be some specificity to this restoring effect and that individual redox regulators have specific targets. Furthermore, thioredoxin also has preferential targets in its reducing activity (e.g., acting much more effectively on Oct-4 than on FoxD3). Therefore, the intracellular redox state may produce different effects on distinct transcription factors that require specific redox partners to restore their activities.

The suggestive picture that arises from these results predicts that Oct-4 is preferentially expressed in the inner cell mass of the blastocyst that will give rise to the embryo. This inner cell mass is characterized by a lower oxygen tension than the surrounding cells, thus enhancing Oct-4 DNA-binding activity.

Since the blastocyst is also the location of thioredoxin requirements during gastrulation (Matsui et al., 1996; Kobayashi-Miura et al., 2002), the redox state of each cell in the blastocyst could define which cell is going to become part of embryonic tissue versus extraembryonic trophoectoderm. Upon uterine implantation and subsequent vascularization, the exposure to increased oxygen tension can down-regulate Oct-4 in the inner cell mass, enabling other less sensitive transcription factors to specify gastrulation and subsequent lineage differentiation.

The charming scenario described suggests a stimulating role of direct electrochemistry in controlling and actuating lineage differentiation in stem cells, allowing one to foresee extremely interesting scenarios in tomorrow's regenerative medicine.

8.3 Role of p53 redox states in DNA binding

p53 is a transcription factor involved in maintaining the integrity of the genome. It responds to DNA damage by promoting cell cycle arrest in G1 phase (Cox & Lane, 1995; Elledge & Lee, 1995; Soussi & May, 1996) and in some cases by inducing apoptosis. In the last two decades, p53 has become one of the most important molecules in the field of cancer research. Indeed, p53 mutants, especially involving mutations in the protein core domain (which is deputed to binding to the p53 consensus sequence; Fojta et al., 1999), have been found in about 50% of human malignancies.

Generally speaking, the diverse functions of p53 are connected to its ability to bind DNA. This ability, in turn, is modulated by the domains of the action protein other than the core one. For example, deletion of the C-terminus gives rise to a p53 that is still active for sequence-specific DNA binding (Hupp et al., 1992; Jayaraman et al., 1997a) and in general the critical role of the basic C-terminus residues in regulating p53-DNA binding has been thoroughly studied. Furthermore, the redox/repair protein Ref-1 (Jayaraman et al., 1997b) is a strong activator of p53 sequence-specific DNA-binding both in vitro and in vivo. From a structural standpoint, it has been shown by X-ray crystallography of the human p53 core domain bound to DNA that the protein contains a zinc ion coordinated by amino acids C176, H179, C238 and C242 (Cho et al., 1994; Arrowsmith & Morin, 1996; Prives 1994). Moreover, chelation of Zn^{2+} abolishes sequence-specific DNA binding by p53 (Srinivasan et al., 1993; Hainaut & Milner, 1993; Hainaut, Butcher, & Milner, 1995).

The fact that the binding of p53 to its DNA consensus sequence is modulated by the redox state of the protein is of particular interest for the ideas presented in this book. Indeed, whereas the reduced protein binds to DNA, the oxidized counterpart appears to hamper irreversibly its ability to bind. This fact is traced back to the loss of the aforementioned Zn ion that is released upon cysteine's oxidation and causes an irreversible structural destabilization in the p53 core domain.

The sensitivity of the p53 DNA-binding activity to its redox state suggests, therefore, the possible efficacy of direct electrochemistry in conditioning its state in order to control the protein's activity in vivo.

8.4 Redox regulation in plants

Redox reactions are essential in plant cell metabolism, being involved in the majority of anabolic and catabolic processes. They take place in many cell sites, such as thylakoid membranes, plastid envelope, plasma membrane and in aqueous cell compartments such as the stroma, the thylakoid lumen and the cytosol in general. Electron-transport systems in cell membranes, i.e., in the photosynthetic and respiratory electron-transport chains, make use of various redox cofactors such as iron-sulfur clusters, quinones and photoexcitable systems that can generate reactive oxygen species. In aqueous-phase cell compartments, the redox state is ruled by metabolites that include NADH, NADPH, glutathione and thiol/disulfide proteins (Foyer & Noctor, 2009). In all the possible electron-transfer reactions that involve the mentioned redox molecules, oxygen can act as electron acceptor, giving rise to potentially harmful reactive oxygen species.

In cells there exists a specialized mechanism that balances redox metabolism and minimizes the formation of reactive oxygen and nitrogen species; this is a redox signaling network that can sense redox imbalance in the environment and react by readjusting redox homeostasis or repairing oxidative damage. Such a network is formed by redox sensors, redox input elements, redox transmitters and redox targets (Dietz, 2008). The basic structure and many components of the thiol-disulfide regulatory network are conserved among cells and cell compartments. In some cases its role is well established, whereas in others its significance is still patchy, due to the ongoing discovery of novel redox targets.

8.4.1 Redox control of plant metabolism

The overall reducing power in plant cells come from the light-induced electron transfer from water to $NADP^+$ that is performed by the photosynthetic apparatus in the thylakoid membranes of chloroplasts. By this process (see section 3.7.3) the electrons are able to cross a total redox potential difference of 1.13 V, which is high enough to power all the subsequent redox reactions in the cell.

The generated reducing power is used by the plant cell mainly in three different types of processes that are interconnected. These are anabolic reactions of metabolism, the antioxidant systems and the redox regulatory system. In metabolism, NADPH is often directly exploited as a cofactor in enzymatic reactions, mainly in anabolic reactions synthesizing molecules of higher complexity or energetic content, for example carbohydrates in the Benson-Calvin cycle or reduced intermediates in sulfur or nitrogen metabolism. Reduced substrates can be used to generate reduction equivalents in the dark or in non-photosynthetic tissues, thereby enabling the plant to uncouple redox-dependent reactions and light reactions of photosynthesis.

Among the aforementioned three types of processes where the reducing power generated by photosynthesis is used, the redox regulatory dithiol/disulfide system is by far the most complex one. It consists of a large number of molecular components that are organized in a hierarchical and highly interconnected network.

Thiol regulation of the activity of Benson-Calvin-cycle enzymes connects light-dependent electron pressure in photosynthetic reactions to ATP and NADPH consumption in reductive carbohydrate metabolism. The regulatory mechanism may be regarded as a prototypical feed-forward activation loop. Indeed, the thiol state-dependent regulation of carbon fluxes through the Benson-Calvin cycle and their link to thioredoxin

(thioredoxin-f)-mediated activation of chloroplast FBPase marked the starting point of long-lasting (over 30 years) research into redox regulation in metabolism (Buchanan & Balmer, 2005). In addition to FBPase, sedoheptulose- 1,7-bisphosphatase, activities of ribulose-5-P kinase, glyceraldehyde-3-P dehydrogenase and rubisco activase are controlled by thioredoxin. Thioredoxin-f donates electrons to target proteins characterized by a broad range of redox midpoint potentials (Hutchison et al., 2000). Differential inactivation of target proteins, e.g., in the Calvin cycle, is not related to the value of midpoint potential but is highly relevant for photoinhibition under non-optimal environmental conditions such as low temperatures (Hutchison et al., 2000). This complexity is partially explained by the fact that thiol modulation is connected to additional metabolic control systems, e.g., the presence of Fru-1,6-BP is needed for FBPase thiol activation (Reichert et al., 2000). Two main carbon pathways drain carbon from the Benson-Calvin cycle, namely Suc synthesis following export of triose phosphate to the cytosol and starch synthesis in the plastids. The committed step of starch synthesis is catalyzed by AGPase. AGPase is activated by reduction of a disulfide bridge between the two slightly smaller subunits of the tetrameric holoenzyme in vitro (Ballicora et al., 2000) and in vivo (Tiessen et al., 2002). Reduction is achieved by thioredoxin-f and thioredoxin-m in vitro and allows for a 4-fold stimulation of ADP-Glc synthesis (Ballicora et al., 2000). A good correlation has been found between Suc concentration, reduction state of the chloroplast and starch synthesis (Tiessen et al., 2002; Geigenberger et al., 2005). NTRC also reductively activates AGPase. NTRC-deficient *Arabidopsis* show less redox-dependent stimulation of AGPase activity and lower starch synthesis rates in high light and upon external feeding of Suc. Inhibition in NTRC knockout plants ranges between 40% and 60% in leaf chloroplasts and reaches 90% in non-photosynthetic amyloplasts (Michalska et al., 2009).

In addition to redox regulation in carbohydrate metabolism, proteomic and biochemical data indicate that thiol modifications also control other major metabolic pathways such as nitrogen assimilation, tetrapyrrole synthesis and lipid synthesis (Lindahl & Kieselbach, 2009).

Lipid synthesis that occurs in the plastids is a strong sink for electrons. Synthesis of palmitic acid (C16) from acetyl-CoA requires 14 molecules of NADPH and seven molecules of ATP. The plastid redox state affects lipid metabolism. Acetyl-CoA carboxylase (ACCase) catalyzes the committed step of malonyl-CoA production in plastid lipid synthesis. Isolated ACCase in vitro is inactive without reductant and activated after addition of DTT or reduced thioredoxin-f or thioredoxin-m (Sasaki et al., 1997). Reductive activation is supported by a pH shift to alkalinization and by increasing Mg^{2+} concentrations.

The chloroplast ACCase consists of four polypeptides, the biotin carboxylase, biotin carboxyl carrier protein, transcarboxylase a-subunit and transcarboxylase b-subunit, with three, one, two and five cysteine residues, respectively (Sasaki et al., 1997). One of the a- or b-subunits is suggested to mediate the redox regulation (Kozaki & Sasaki, 1999). The biotin carboxyl carrier subunit of ACCase in *Chlamydomonas reinhardtii* is subjected to *S*-thiolation with glutathione (Michelet et al., 2008). Biotin carboxylase is the target of glutathionylation in *Arabidopsis* cell culture (Dixon et al., 2005). Thus, each of the subunits of ACCase is potentially controlled by redox regulation using diverse mechanisms.

This fact underlines the connection between redox state and lipid metabolism. Envelope-bound monogalactosyldiacylglycerol synthase (MGD) synthesizes monogalactosyldiacylglycerol from diacylglycerol and UDP-Gal. Monogalactosyldiacylglycerol is a major lipid component of chloroplasts. In vitro MGD activity that depends on the presence of reductants (e.g., DTT) is inhibited by thiol-alkylating agents, and is modulated by thioredoxin acting on intramolecular disulfide bonds (Yamaryo et al., 2006). Plant MGD possesses nine conserved cysteine residues. Its regulation by the thiol redox state is thought to enable galactolipid synthesis along with photosynthetic activity and to foster replacement of oxidized lipids under conditions that cause oxidative stress (Yamaryo et al., 2006).

8.4.2 Redox regulation of gene transcription in plastids

Plastids possess their own genome, the plastome, and a complete molecular machinery to express the genetic information in it. Although this plastome encodes approximately just 120 genes (in vascular plants) the expression mechanisms appear to be rather complex and highly regulated. This includes a number of redox control mechanisms that influence regulatory proteins at any important level of gene expression, i.e., transcription, posttranscriptional modification and translation initiation (Pfannschmidt & Liere, 2005).

A major target of photosynthetic redox signals is the plastid-encoded RNA polymerase (PEP). Unbalanced excitation of the two photosystems generates either a reduced or oxidized pool of PQ that act as signals that control the phosphorylation of the light-harvesting complexes of PSII via the thylakoid-associated kinase STN7 (Lemeille & Rochaix, 2010; Pesaresi et al., 2010).

The same signals also trigger a phosphorylation cascade towards the PEP enzyme that results in alterations of photosynthesis gene expression

(Allen & Pfannschmidt, 2000). Both processes have the effect of counteracting the unbalanced excitation and maintaining the highest photosynthesis efficiency. The phosphorylation cascade probably includes the action of a number of further kinases (STN8, an ortholog of STN7; CSK, the chloroplast sensor kinase; PTK, the plastid transcription kinase), generating a phosphorylation network. In a simplified view, the reduction of the PQ pool activates STN7, which provides an input signal for the subsequent kinase network. This controls the phosphorylation state of the sigma factor Sig1 that in turn regulates the relative transcription of the photosynthesis reaction center genes psbA (encoding the D1 protein of PSII) and psaA/B (encoding the P700 apoprotein of PSI; Shimizu et al., 2010). This view is consistent with the observation that CSK, PTK and Sig1 are able to interact with each other in the yeast (*Saccharomyces cerevisiae*) two-hybrid system (Puthiyaveetil et al., 2010). In organello transcription experiments in the presence of kinase inhibitors and/or the reductant DTT, however, indicate that this phosphorylation-dependent signal interacts with a second, thiol-dependent signal (Steiner et al. 2009). PTK, a casein-kinase 2 type enzyme, has been reported to be under control of the redox state of glutathione (Ogrzewalla et al., 2002), but its activity could not be modulated with DTT. This suggested the involvement of a further regulator.

Recently, two independent studies identified a novel thioredoxin-like protein that probably represents the sought-for additional player (Arsova et al., 2010; Schröter et al., 2010). The novel thioredoxin was named thioredoxin-z because of its distinct evolutionary position in relation to thioredoxin-x and thioredoxin-y. In a yeast two-hybrid screen it has been identified as an interacting protein of two chloroplast-located phosphofructokinase-like proteins called FLN1 and FLN2. The *Arabidopsis* knockout mutant line of thioredoxin-z exhibits pale-white leaves and is viable only on Suc-supplemented medium, a unique phenotype since the thioredoxin system is highly redundant and can easily compensate for the loss of single components. Gene expression analyses indicated the same plastid gene expression profiles as in PEP-deficient mutants, pointing to an important role of thioredoxin-z in plastid development and gene expression (Arsova et al., 2010). These observations have been complemented by mass spectrometry results that demonstrate that both the thioredoxin-like protein and the FLN2 kinase are intrinsic subunits of the PEP enzyme of chloroplasts (Schröter et al., 2010). This provides a direct explanation for the phenotype and the expression profiles in the knockout mutant. A lack of thioredoxin-z prevents proper assembly of the PEP enzyme and, as a consequence, the developmental transition from the nuclear-encoded RNA polymerase-driven transcription to the PEP-dependent transcription does not take place. The precise functional role of thioredoxin-z within the PEP complex, its relation to FLN1 and FLN2, as well as its regulatory impact remain to be elucidated. Furthermore, it is still a matter of debate how it relates to the other known redox regulators mentioned above. In summary, our understanding of photosynthetic redox signal transduction towards the level of gene expression is still at the beginning. The increasing number of identified regulatory components unravels a complex mechanistic redox tool-box enabling chloroplasts to respond to a wide range of environmental conditions in a dynamic and flexible manner. The elucidation of the role of the regulatory components will help in selecting the pathways that will be more prone to be directly under external electrochemical control.

8.5 The electrified snail

We conclude the review of possible biological systems that appear to be potentially suitable to direct electrochemical control with a beautiful example (Halámková et al., 2012) that involves many of the concepts we have introduced in the chapters of this book. It demonstrates a bio-fuel cell directly implanted on a living snail that is fueled by the glucose the snail gets with its food. This is the first example of time-sustainable generation of electrical power in vivo and shows that metabolic activity can "recharge" the living battery for continuous production of electricity.

The critical input provided on the one hand by a deep knowledge of bioelectrochemistry and surface biofunctionalization, and the potential expressed by the involvement of nanotechnologies on the other hand, allowed the setting-up of a biocompatible, efficient and robust electrode pair made of compressed carbon nanotubes modified with PBSE linker (a heterobifunctional cross-linker, 1-pyrenebutanoic acid succinimidyl ester) and then functionalized with laccase or PQQ-GDH (pyrroloquinoline quinone (PQQ)-dependent glucose dehydrogenase) to yield the biocatalytic cathode or anode, respectively. The implantable electrodes were inserted into the snail through two holes cut in the shell and placed into the hemolymph (i.e., the snail's blood) between the body wall and internal organs (visceral mass). The implanted electrodes performing the bioelectrocatalytic reactions (glucose oxidation at the anode and O_2 reduction at the cathode) were connected to a circuit composed of an external variable-load resistance; voltage and current were measured during biofuel cell operation in vivo.

The open-circuit voltage and short-circuit current achieved in the biofuel cell were 530 mV and 42.5 μA (current density 170 $\mu A\ cm^{-2}$), respectively.

The sustainability of the implanted biofuel cell was tested by measuring the voltage and current produced over time on the optimum load resistance of 20 kΩ (corresponding to the internal resistance of the cell and providing a maximum power of 7.45 μW (i.e., a power density of about 30 μW cm^{-2}). It turned out that the electrical output decreased rapidly upon cell operation; however, it was effectively restored when the current extraction was interrupted for 30–60 min to allow the snail to rest. The biofuel cell operation was reproducible even after a period of 2 weeks and was not affected by enzyme inactivation and/or biofouling due to the biological environment.

This experiment demonstrates clearly the state of development of the enabling technology that is needed for connecting living beings to an external electrical circuit. Its success represents an encouraging starting point for future, more sophisticated implementations of direct electrochemical control of biological systems and reactions in vivo. We firmly believe that we are at the beginning of a new, charming adventure in human technology that will soon provide results that are, to date, still far beyond our imagination.

8.6 References

Akamatsu, Y., Ohno, T., Hirota, K., Kagoshima, H., Yodoi, J., & Shigesada, K. (1997). Redox regulation of the DNA binding activity in transcription factor PEBP2: the roles of two conserved cysteine residues. *J. Biol. Chem., 272*, 14497–14500.

Allen, J. F., & Pfannschmidt, T. (2000). Balancing the two photosystems: photosynthetic electron transfer governs transcription of reaction centre genes in chloroplasts. *Phil. Trans. R. Soc. Lond. B Biol. Sci., 355*, 1351–1359.

Arrowsmith, C. H., & Morin, P. (1996). New insights into p53 function from structural studies. *Oncogene, 12*, 1379–1385.

Arsova, B., Hoja, U., Wimmelbacher, M., Greiner, E., Ustün, S., Melzer, M., Petersen, K., Lein, W., & Börnke, F. (2010). Plastidial thioredoxin z interacts with two fructokinase-like proteins in a thiol-dependent manner: evidence for an essential role in chloroplast development in *Arabidopsis* and Nicotiana benthamiana. *Plant Cell, 22*, 1498–1515.

Ballicora, M. A., Frueauf, J. B., Fu, Y. B., Schürmann, P., & Preiss, J. (2000). Activation of the potato tuber ADP-glucose pyrophosphorylase by thioredoxin. *J. Biol. Chem., 275*, 1315–1320.

Buchanan, B. B., & Balmer, Y. (2005). Redox regulation: a broadening horizon. *Annu. Rev. Plant. Biol., 56*, 187–220.

Cho, Y., Gorina, S., Jeffrey, P. D., & Pavletich, N. P. (1994). Crystal structure of a p53 tumor suppressor-DNA complex: understanding tumorigenic mutations. *Science, 265*, 346–355.

Cox, L. S., & Lane, D. P. (1995). Tumour suppressors, kinases and clamps: how p53 regulates the cell cycle in response to DNA damage. *Bioessays, 17*, 501–508.

Dietz, K. J. (2008). Redox signal integration: from stimulus to networks and genes. *Physiol. Plant, 133*, 459–468.

Dixon, D. P., Skipsey, M., Grundy, N. M., & Edwards, R. (2005). Stress-induced protein S-glutathionylation in. *Arabidopsis. Plant. Physiol., 138*, 2233–2244.

Elledge, R. M., & Lee, W.-H. (1995). Life and death by P53. *Bioessays, 17*, 923–930.

Fojta, M., Kubicárová, T., Vojtesek, B., & Palecek, E. (1999). Effect of p53 protein redox states on binding to supercoiled and linear DNA. *J. Biol. Chem., 274*, 25749–25755.

Foyer, C. H., & Noctor, G. (2009). Redox regulation in photosynthetic organisms: signaling, acclimation, and practical implications. *Antioxid. Redox Signal., 11*, 861–905.

Geigenberger, P., Kolbe, A., & Tiessen, A. (2005). Redox regulation of carbon storage and partitioning in response to light and sugars. *J. Exp. Bot., 56*, 1469–1479.

Guo, Y., Einhorn, L., Kelley, M., Hirota, K., Yodoi, J., Reinbold, R., Scholer, H., Ramsey, H., & Hromas, R. (2004). Redox regulation of the embryonic stem cell transcription factor oct-4 by thioredoxin. *Stem Cells, 22*, 259–264.

Hainaut, P., & Milner, J. (1993). A structural role for metal ions in the wild-type conformation of the tumor suppressor protein p53. *Cancer Res., 53*, 1739–1742.

Hainaut, P., Butcher, S., & Milner, J. (1995). Temperature sensitivity for conformation is an intrinsic property of wild-type p53. *Br. J. Cancer, 71*, 227–231.

Halamkova, L., Halamek, J., Bocharova, V., Szczupak, A., Alfonta, L., & Katz, E. (2012). Implanted biofuel cell operating in a living snail. *J. Am. Chem. Soc., 134*, 5040–5043.

Hirota, K., Matsui, M., Iwata, S., Nishiyama, A., Mori, K., & Yodoi, J. (1997). AP-1 transcriptional activity is regulated by a direct association between thioredoxin and Ref-1. *Proc. Natl. Acad. Sci. USA, 94*, 3633–3638.

Hirota, K., Murata, M., Sachi, Y., Nakamura, H., Takeuchi, J., Mori, K., & Yodoi, J. (1999). Distinct roles of thioredoxin in the cytoplasm and in the nucleus: a two-step mechanism of redox regulation of transcription factor NF-κB. *J. Biol. Chem., 274*, 27891–27897.

Hupp, T. R., Meek, D. W., Midgley, C. A., & Lane, D. P. (1992). Regulation of the specific DNA binding function of p53. *Cell, 71*, 875–886.

Hutchison, R. S., Groom, Q., & Ort, D. R. (2000). Differential effects of chilling induced photooxidation on the redox regulation of photosynthetic enzymes. *Biochemistry, 39*, 6679–6688.

Jayaraman, L., Freulich, E., & Prives, C. (1997a). Functional dissection of p53 tumor suppressor protein. *Methods Enzymol., 283*, 245–226.

Jayaraman, L., Murthy, K. G., Zhu, C., Curran, T., Xanthoudakis, S., & Prives, C. (1997b). Identification of redox/repair protein Ref-1 as a potent activator of p53. *Genes Dev., 11*, 558–570.

Kobayashi-Miura, M., Nakamura, H., Yodoi, J., & Shiota, K. (2002). Thioredoxin, an anti-oxidant protein, protects mouse embryos from oxidative stress-induced developmental anomalies. *Free Radic. Res., 36*, 949–956.

Kozaki, A., & Sasaki, Y. (1999). Light-dependent changes in redox status of the plastidic acetyl-CoA carboxylase and its regulatory component. *Biochem. J., 339*, 541–546.

Lemeille, S., & Rochaix, J. D. (2010). State transitions at the crossroad of thylakoid signalling pathways. *Photosynth. Res., 106*, 33–46.

Lindahl, M., & Kieselbach, T. (2009). Disulphide proteomes and interactions with thioredoxin on the track towards understanding redox regulation in chloroplasts and cyanobacteria. *J. Proteomics, 72*, 416–438.

Matsui, M., Oshima, M., Oshima, H., Takaku, K., Maruyama, T., Yodoi, J., & Taketo, M. M. (1996). Early embryonic lethality caused by targeted disruption of the mouse thioredoxin gene. *Dev. Biol., 178*, 179–185.

Michalska, J., Zauber, H., Buchanan, B. B., Cejudo, F. J., & Geigenberger, P. (2009). NTRC links built-in thioredoxin to light and sucrose in regulating starch synthesis in chloroplasts and amyloplasts. *Proc. Natl. Acad. Sci. USA, 106*, 9908–9913.

Michelet, L., Zaffagnini, M., Vanacker, H., Le Maréchal, P., Marchand, C., Schroda, M., Lemaire, S. D., & Decottignies, P. (2008). In vivo targets of S-thiolation in *Chlamydomonas reinhardtii*. *J. Biol. Chem., 283*, 21571–21578.

Ogrzewalla, K., Piotrowski, M., Reinbothe, S., & Link, G. (2002). The plastid transcription kinase from mustard (*Sinapis alba* L.): a nuclear-encoded CK2-type chloroplast enzyme with redox-sensitive function. *Eur. J. Biochem., 269*, 3329–3337.

Pesaresi, P., Hertle, A., Pribi, M., Schneider, A., Kleine, T., & Leister, D. (2010). Optimizing photosynthesis under fluctuating light: the role of the *Arabidopsis* STN7 kinase. *Plant Signal Behav., 5*, 21–25.

Pfannschmidt, T., & Liere, K. (2005). Redox regulation and modification of proteins controlling chloroplast gene expression. *Antioxid. Redox Signal, 7*, 607–618.

Prives, C. (1994). How loops, beta sheets, and alpha helices help us to understand p53. *Cell, 78*, 543–546.

Puthiyaveetil, S., Ibrahim, I. M., Jelicic, B., Tomasic, A., Fulgosi, H., & Allen, J. F. (2010). Transcriptional control of photosynthesis genes: the evolutionarily conserved regulatory mechanism in plastid genome function. *Genome Biol. Evol., 2*, 888–896.

Rajagopalan, S., Kurz, S., Münzel, T., Tarpey, M., Freeman, B. A., Griendling, K. K., & Harrison, D. G. (1996). Angiotensin II-mediated hypertension in the rat increases vascular superoxide production via membrane NADH/NADPH oxidase activation. Contribution to alterations of vasomotor tone. *J. Clin. Invest., 97*, 1916–1923.

Reichert, A., Baalmann, E., Vetter, S., Backhausen, J. E., & Scheibe, R. (2000). Activation properties of the redox-modulated chloroplast enzymes glyceraldehyde

3-phosphate dehydrogenase and fructose-1,6-bisphosphatase. *Physiol. Plant, 110*, 330–341.

Sasaki, Y., Kozaki, A., & Hatano, M. (1997). Link between light and fatty acid synthesis: thioredoxin-linked reductive activation of plastidic acetyl-CoA carboxylase. *Proc. Natl. Acad. Sci. USA, 94*, 11096–11101.

Schröter, Y., Steiner, S., Matthäi, K., & Pfannschmidt, T. (2010). Analysis of oligomeric protein complexes in the chloroplast sub-proteome of nucleic acid-binding proteins from mustard reveals potential redox regulators of plastid gene expression. *Proteomics, 10*, 2191–2204.

Shimizu, M., Kato, H., Ogawa, T., Kurachi, A., Nakagawa, Y., & Kobayashi, H. (2010). Sigma factor phosphorylation in the photosynthetic control of photosystem stoichiometry. *Proc. Natl. Acad. Sci. USA, 107*, 10760–10764.

Soussi, T., & May, P. (1996). Structural aspects of the p53 protein in relation to gene evolution: a second look. *J. Mol. Biol., 260*, 623–637.

Srinivasan, R., Roth, J. A., & Maxwell, S. A. (1993). Sequence-specific interaction of a conformational domain of p53 with DNA. *Cancer Res., 53*, 5361–5364.

Steiner, S., Dietzel, L., Schröter, Y., Fey, V., Wagner, R., & Pfannschmidt, T. (2009). The role of phosphorylation in redox regulation of photosynthesis genes psaA and psbA during photosynthetic acclimation of mustard. *Mol. Plant, 2*, 416–429.

Tiessen, A., Hendriks, J. H. M., Stitt, M., Branscheid, A., Gibon, Y., Farré, E. M., & Geigenberger, P. (2002). Starch synthesis in potato tubers is regulated by post-translational redox modification of ADP-glucose pyrophosphorylase: a novel regulatory mechanism linking starch synthesis to the sucrose supply. *Plant Cell, 14*, 2191–2213.

Ueno, M., Masutani, H., Arai, R., Yamauchi, A., Hirota, K., Sakai, T., Inamoto, T., Yamaoka, Y., Yodoi, J., & Nikaido, T. (1999). Thioredoxin-dependent redox regulation of p53-mediated p21 activation. *J. Biol. Chem., 274*, 35809–35815.

Yamaryo, Y., Motohashi, K., Takamiya, K., Hisabori, T., & Ohta, H. (2006). In vitro reconstitution of monogalactosyldiacylglycerol (MGDG) synthase regulation by thioredoxin. *FEBS Lett., 580*, 4086–4090.

Zhou, A., Carrell, R. W., Murphy, M. P., Wei, Z., Yan, Y., Stanley, P. L. D., Stein, P. E., Broughton Pipkin, F., & Read, R. J. (2010). A redox switch in angiotensinogen modulates angiotensin release. *Nature, 468*, 108–111.

Index

Note: Page numbers with "*f*" denote figures; "*t*" tables.

P

R

Kolibri 360 GmbH
Ferdinand-Jühlke-Straße 7
99095 Erfurt, Deutschland
produktsicherheit@kolibri360.de